The Handbook of Environmental Chemistry

Volume 1 Part F

Edited by O. Hutzinger

The Handbook
of Environmental Chemistry

Volume 2 Part F

The Natural Environment and the Biogeochemical Cycles

With contributions by
W. S. Fyfe, H. Puchelt, M. Taube

With 42 Figures and 54 Tables

Springer-Verlag Berlin Heidelberg GmbH

Professor Dr. Otto Hutzinger

University of Bayreuth
Chair of Ecological Chemistry and Geochemistry
P.O. Box 101251, W-8580 Bayreuth, Federal Republic of Germany

ISBN 978-3-662-14985-0   ISBN 978-3-540-46995-7 (eBook)
DOI 10.1007/978-3-540-46995-7

Library of Congress Cataloging-in-Publication Data
(Revised for volume F) The Natural environment and the biogeochemical cycles. (The Handbook of environmental
chemistry; v. 1, part A-) Includes bibliographical references and indexes.
1. Biogeochemical cycles--Collected works.
2. Environmental chemistry--Collected works.
I. Craig, P. J., 1944-. II. Series: Handbook of environmental chemistry; v. 1, pt. A, etc. QD31.H335 vol. 1, pt. A, etc.
628.5 s 80-16608 [QH344] [574.5′222]

© Springer-Verlag Berlin Heidelberg 1992
Originally published by Springer-Verlag Berlin Heidelberg New York in 1992.
Softcover reprint of the hardcover 1st edition 1992

Typesetting: Macmillan India Ltd., Bangalore-25, India

52/3020-5 4 3 2 1 0-Printed on acid-free paper

Preface

Environmental Chemistry is a relatively young science. Interest in this subject, however, is growing very rapidly and, although no agreement has been reached as yet about the exact content and limits of this interdisciplinary subject, there appears to be increasing interest in seeing environmental topics which are based on chemistry embodied in this subject. One of the first objectives of Environmental Chemistry must be the study of the environment and of natural chemical processes which occur in the environment. A major purpose of this series on Environmental Chemistry, therefore, is to present a reasonably uniform view of various aspects of the chemistry of the environment and chemical reactions occurring in the environment.

The industrial activities of man have given a new dimension to Environmental Chemistry. We have now synthesized and described over five million chemical compounds and chemical industry produces about one hundred and fifty million tons of synthetic chemicals annually. We ship billions of tons of oil per year and through mining operations and other geophysical modifications, large quantities of inorganic and organic materials are released from their natural deposits. Cities and metropolitan areas of up to 15 million inhabitants produce large quantities of waste in relatively small and confined areas. Much of the chemical products and waste products of modern society are released into the environment either during production, storage, transport, use or ultimate disposal. These released materials participate in natural cycles and reactions and frequently lead to interference and disturbance of natural systems.

Environmental Chemistry is concerned with *reactions in the environment*. It is about distribution and equilibria between environmental compartments. It is about reactions, pathways, thermodynamics and kinetics. An important purpose of this Handbook is to aid understanding of the basic distribution and chemical reaction processes which occur in the environment.

Laws regulating toxic substances in various countries are designed to assess and control risk of chemicals to man and his environment. Science can contribute in two areas to this assessment: firstly in the area of toxicology and secondly in the area of chemical exposure. The available concentration ("environmental exposure concentration") depends on the fate of chemical compounds in the environment and thus their distribution and reaction behaviour in the environment. One very important contribution of Environmental Chemistry to the above mentioned toxic substances laws is to develop laboratory test methods, or mathematical correlations and models that predict the environmental fate of new chemical compounds. The third purpose of this Handbook is to help in the basic understanding and development of such test and models.

The last explicit purpose of the handbook is to present, in a concise form, the most important properties relating to environmental chemistry and hazard assessment for the most important series of chemical compounds.

At the moment three volumes of the Handbook are planned. Volume 1 deals with the natural environment and the biogeochemical cycles therein, including some background information such as energetics and ecology. Volume 2 is concerned with reactions and processes in the environment and deals with physical factors such as transport and adsorption, and chemical, photochemical and biochemical reactions in the environment, as well as some aspects of pharmacokinetics and metabolism within organisms. Volume 3 deals with anthropogenic compounds, their chemical backgrounds, production methods and information about their use, their environmental behaviour, analytical methodology and some important aspects of their toxic effects. The material for volumes 1, 2, and 3 was more than could easily be fitted into a single volume, and for this reason, as well as for the purpose of rapid publication of available manuscripts, all three volumes are published as a volume series (e.g. Vol. 1; A, B, C). Publisher and editor hope to keep the material of the volumes 1 to 3 up to date and to extend coverage in the subject areas by publishing further parts in the future. Readers are encouraged to offer suggestions and advice as to future editions of "The Handbook of Experimental Chemistry".

Most chapters in the Handbook are written to a fairly advanced level and should be of interest to the graduate student and practising scientist. I also hope that the subject matter treated will be of interest to people outside chemistry and to scientists in industry as well as government and regulatory bodies. It would be very satisfying for me to see the books used as a basis for developing graduate courses on Environmental Chemistry.

Due to the breadth of the subject matter, it was not easy to edit this Handbook. Specialists had to be found in quite different areas of science who were willing to contribute a chapter within the prescribed schedule. It is with great satisfaction that I thank all authors for their understanding and for devoting their time to this effort. Special thanks are due to the Springer publishing house and finally I would like to thank my family, students and colleagues for being so patient with me during several critical phases of preparation for the Handbook, and also to some colleagues and the secretaries for their technical help.

I consider it a privilege to see my chosen subject grow. My interest in Environmental Chemistry dates back to my early college days in Vienna. I received significant impulses during my postdoctoral period at the University of California and my interest slowly developed during my time with the National Research Council of Canada, before I was able to devote my full time to

Environmental Chemistry in Amsterdam. I hope this Handbook will help deepen the interest of other scientists in this subject.

Otto Hutzinger

This preface was written in 1980. Since then publisher and editor have agreed to expand the Handbook by two new open-ended volume series: Air Pollution and Water Pollution. These broad topics could not be fitted easily into the headings of the first three volumes.

All five volume series will be integrated through the choice of topics covered and by a system of cross referencing.

The outline of the Handbook is thus as follows:

1. The Natural Environment and the Biogeochemical Cycles
2. Reactions and Processes
3. Anthropogenic Compounds
4. Air Pollution
5. Water Pollution

Bayreuth, March 1992   Otto Hutzinger

Contents

W. S. Fyfe   Geosphere Interactions on a Convecting Planet:
Mixing and Separation   1

H. Puchelt   Environmental Inorganic Chemistry of the
Continental Crust   27

M. Taube   Evolution of Matter and Energy   65

Subject Index   183

Volume Index   185

List of Contributors

Prof. Dr. W. S. Fyfe
University of Western Ontario
Department of Geology
London, Ontario N6A 5B7
Canada

Prof. Dr. Mieczyslaw Taube
Bollackerweg 10
CH-8956 Killwangen

Prof. Dr. Harald Puchelt
Institut für Petrographie
und Geochemie
Universität Karlsruhe
Kaiserstraße 12
7500 Karlsruhe
FRG

Geosphere Interactions On A Convecting Planet: Mixing and Separation

W.S. Fyfe

Faculty of Science, University of Western Ontario,
London, Ontario, Canada N6A 5B7

Introduction   3
A Convecting Planet   4
The Present Convective Style   5
Chemical Separation on and in the Earth   7
Chemical Processes at Ridges: Mixing of
   Crust-Hydrosphere-Atmosphere   7
Ridge – Summary 12
The Quiet Ocean Floor 12
Subduction: Return Flow to the Mantle 14
Continental Collisions and Global Geochemistry 19
Hot Spots 20
Mixing and Tectonics 21
The Creation of High Topography 21
Atmosphere-Hydrosphere-Crust-Biosphere Interactions 22
Deep Recycling 24
References 25

Summary

All parts of the plate tectonic cycle of Earth are associated with major geochemical separation and mixing processes. Ocean ridge processes, which move about half the internal energy to the surface, involve the separation of basaltic magmas from the upper mantle. At the ridges, cooling by ocean water, which removes about half the energy from the hot rocks, leads to massive chemical modification of oceans and the upper layers of the oceanic crust. The crust gains volatiles (H_2O, CO_2, N_2, Cl, S) and species like Na, K, Mg, U. The oceans gain important quantities of species such as Mn, Fe, Ca, Ni, Cu, Zn and major additions of SiO_2, Ca, Li. The ocean strontium isotope ratio dominated by continental runoff (0.711) is reduced to 0.709 by deep flows. The cooling fluxes process the entire ocean mass in about 2 million years. Little is yet known about processes on the quiet ocean floors which cover 60% of the planet. But present evidence suggests that convective flow beneath the sediment cover continues all the way to subduction. Evidence is increasing for massive serpentinization processes in deep layers of the ocean crust, a process which may contribute to heat production and formation of very saline fluids.

The Handbook of Environmental Chemistry
Volume 1 Part F, Ed. O. Hutzinger
© Springer-Verlag Berlin Heidelberg 1992

At subduction zones, the volatile-laden ocean crust, with a component of sediments, which may depend on the roughness of the surface of the oceanic lithosphere, is essentially quantitatively subducted. This process recharges the mantle with volatiles and species like Na, K, U, and carbon and sulphur compounds. Release of the volatiles must induce convection in the overlying mantle and mobilize the most soluble and fusible components eventually leading to the production of basaltic-andesites. In most subduction zones, such processes occur beneath the continental crust. The scale of the process indicates that the ocean mass is recycled through the mantle in about a billion years. Magma underplating processes above subduction zones lead to massive reworking of the continental crust and to a steepened thermal gradient. Granite-rhyolite production and prograde metamorphism essentially dehydrates the basal crust and leads to a host of ore-forming processes. At high crustal levels, water cooling processes again transfer heat and soluble components as observed in geothermal systems.

Similar processes occur during continental collisions (and on a small scale with strike-slip boundaries) where crust may be thickened to 60–80 km. In the Himàlayan example, a vast metamorphic dewatering and melting event follows the major thrust events. The mass of metamorphic fluids in this case is similar to that of the ice caps. Where hydrocarbon-salt basins are involved in such events, massive remobilization and reconcentration may occur.

All mantle processes, ridges, hot spots, subduction-collisions create the high topography which focus the interactions with surface fluids. Weathering by solution and particle transport is one of Earth's great chemical-tectonic processes and is the final relaxation process of the mantle derived tectonic events.

Modern seismic and deep electric remote sensing techniques, and deep drilling, increasingly confirm the scale of all such processes which involve fluids. At the present time, almost one third of the continental crust is involved in such heat-mass transport processes. We live on a water cooled planet and all these processes support the biomass of Earth.

Introduction

Since the time of the great geochemists of the early part of this century (e.g. Goldschmidt, Vernadsky, Correns) it has been traditional to treat the geochemistry of planet Earth in terms of its separate parts, the geospheres. The situation is well illustrated by Mason's classic textbook of 1966 [1].

Most present models of the Earth [2] divide it into:

a) The inner core, solid and dense ($13\ \mathrm{g\,cm^{-3}}$) dominated by crystalline metal.
b) The outer core, liquid and dense ($11\text{--}12\ \mathrm{g\,cm^{-3}}$) dominated by liquid metal.
c) The lower mantle, solid, density near $5\ \mathrm{g\,cm^{-3}}$, dominated by Mg–Fe–Si–O.
d) The upper mantle, solid, density $3.5\text{--}4.0\ \mathrm{g\,cm^{-3}}$, dominated by Mg–Fe–Si–O.
e) The asthenosphere, essentially solid but perhaps with a small molten component, dominated by Mg–Fe–Si–O.
f) The lithosphere, a crystalline outer layer $100\text{--}300$ km thick, solid and carrying the crust of the earth with complex chemistry (the low melting fraction of the planet).
g) The hydrosphere, including all the systems of earth dominated by liquid water.
h) The atmosphere, dominated by oxygen and nitrogen with many other gases in small quantities (e.g. ozone, methane, water, carbon dioxide).
i) The biosphere-living matter dominated by C–O–H and various crystalline and amorphous included matter. More recently there is a tendency to see the use of the word ecosphere – the total system of minerals-H_2O-atmosphere-life, which contains and envelops all living matter.

This gross earth model was built on observations of the accessible parts plus indirect observations of the deeper interior using data from studies of seismic waves through the planet. The mean density, mass, and moment of inertia of the planet, the chemistry of meteorites, studies of the states of matter over the pressure-temperature range of the planet also aided its development. Such models have been continuously refined with the increasing sophistication of seismic methods, global seismic networks and the use of large computers to analyze data. The new "global tomography" [2, 3] increasingly shows detail of the main boundary regions of earth. Today this is being amplified by improved electrical methods [4] which provide information on the resistivity of earth materials to depths of several hundred kilometers. But while models improve in detail, and mantle structures grew more complicated, the gross picture established decades ago is still essentially valid. What has changed dramatically over the past two decades is our knowledge of earth dynamics; how the planet changes with time and the interactions between the major geospheres. While recycling of materials in the near surface was widely recognized in the 1950s, the total picture has changed dramatically with greatly increased emphasis on significant recycling of materials at least to upper mantle depths. Also, the new observations on other planets in the solar system have sharpened our views on planetary evolution in the solar system.

A Convecting Planet

Over the past decades new observations of Earth, particularly those from the ocean floors, have shown that Earth is an actively convecting body. Previously, many concerned with the physical properties of the deeper parts had considered that some properties, particularly viscosity, were such that Rayleigh numbers of materials at depth would be too great to allow significant convection. Such considerations led to the great debate over "continental drift", possible or impossible?

Modern observations have resolved this debate. Data on other planets and in particular the Moon, Mars, Venus [2] indicate that the planets were rapidly heated during their accumulation from the dust cloud of the solar system. Planets were almost totally assembled in something like 100 million years after the birth of the solar system. Gravitational collapse would provide enough energy to almost totally melt a body the size of the Earth. Energy would also be supplied by a host of short-lived radioactive isotopes in the early solar dust cloud. Separation of the Moon from the Earth may have resulted from a major impact [2]. Most of the larger planets could have been covered by magma (molten) oceans. With such a scenario, heavy materials, in particular liquid or solid metals would rapidly sink to form the core, liberating gravitational energy. A heavy atmosphere rich in volatiles ($H_2O–CO_2–Ar–N_2$) would form in equilibrium with the surface molten zone of the planet.

We know that the planetary objects in the solar system (and many of the meteorites) formed about 4.6 billion year ago. The oldest materials on this planet (some zircon crystals from Australia) are slightly over 4 billion years old. The oldest rock masses or any significant scale are preserved from about 4.0 billion years ago [5]. The record of the first 500 million years is essentially missing, a tribute to the turbulence of the early planetary surface. But what is perhaps surprising, is that the oldest rocks, with a few exceptions, are really quite similar in type to those of the modern Earth even if the quantitative relations and geologic structures are a little different [5–6].

The Earth system today, Earth dynamics, is powered by a number of heat sources. First, the surface environment is dominated by solar heating, which because of our atmosphere and albedo, maintains surface water in the liquid state and all present evidence shows that liquid water has been present for four billion years! As we shall see below, this has a profound effect on the dynamics and chemical evolution of Earth. Heat flows from the interior and this energy is supplied by:

a) residual energy from accumulation which perhaps mainly resides in the thermostat of the solid-liquid, essentially metallic core system from which the latent heat of crystallization will be slowly released, and

b) the radioactive decay of elements with long half-lives and in particular isotopes of potassium, uranium and thorium in the mantle. Most earth models consider that these sources can explain the heat flow observed at the modern surface [2].

Earth can lose energy from the interior by the two major processes of conduction and convection. Slowly, we are beginning to understand the balance between the two processes. Active convection requires an appropriate Rayleigh number [7] for the layers of the interior and modern analysis of the problem shows that the Ra's for the interior in general greatly exceed those required for the initiation of convection. While in general, pressure increases viscosity, new data from high-pressure experimentation and seismic modelling, has shown that the expected increase in viscosity is offset by phase changes and changes in coordination numbers as simple crystal structures dominate materials at depth. In fact viscosity changes little through the outer 100s of km of the mantle.

For any regular object cooling from its surface, it is not difficult to determine if convection is occurring. For a uniform material cooling by conduction, heat flow (q = K grad T) should be uniform over the surface. If convection occurs, the coupled heat-mass transport will cause irregularity in the surface heat flow. Further, horizontal thermal irregularity in the body will cause an irregular topography of the surface. The most casual inspection of the topography of the solid surface of Earth shows gross irregularity and further, the surface topography can be correlated with the expected irregularity in heat flow for a convecting body. The Earth is cooling by conduction and convection and modern data shows that the total energy flow is transmitted about equally by the two processes at the present time. Consideration of the thermal history of a cooling object also shows, that in the past, convection must have been more important or the dominant process (for the present Moon and Mars, heat loss appears to be essentially by conduction only). We now are beginning to appreciate that the convection processes occuring on Earth drive the surface processes and are essential to the maintenance of the biosphere or ecosphere and the environment. Further, fluctuations in the convective system play a crucial role in changes of the surface environment. One of the remarkable features of our planet is that while there have been major fluctuations, to the best of our knowledge, Earth's hydrosphere has never totally frozen and never boiled [8]! The system which supports life has been maintained more or less constant for 4.0 billion years.

The Present Convective Style

Convective mass-energy flow from the interior occurs dominantly at the great narrow ocean ridge systems of the Earth (the mid-Atlantic ridge, Indian Ocean Ridge, East Pacific Rise, etc.), which are dominantly submarine and form the most continuous mountain ranges of Earth. Some have dimensions of almost half the global circumference.

These submarine mountains chains are volcanic with a surface expression of basaltic lava flows. From all the Earth's ridges about 15 km^3 of magma is produced annually by melting in rising hot mantle plumes through pressure release. About 90% of our planet's volcanism occurs at these submarine sites.

Ridge volcanism accounts for almost half the heat loss from the planet's interior. The rising convection cells root at depths of several hundred kilometres. Magma batches that rise into the ridges may reach tens of km^3 in volume at any time [9, 10]. Active magma chambers have only recently been mapped and single eruptions of 15 km^3 have been observed [11]. The rocks of the ridges produce a crystalline product of essentially zero age and produce the new crust of the Earth with typical basaltic composition. While there are small differences, this composition rarely departs from the mean for any species by more than a few percent. The basalt represents the low melting point fracture of the upper mantle beneath oceanic crust. As the basaltic material penetrates to the surface, it pushes the crust apart and causes the process we now call ocean floor spreading, the driving force of continental drift. The process brings to the surface a new supply of many elements essential to the nutrient supply of the biosphere.

Early studies of heat flow over and near the ridges produced a great surprise. Given that the rising magmas which extrude at and intrude below the surface at temperatures in excess of 1200 °C, it would be expected that very high heat flow would be observed. While many high heat flow values were observed, many others were much less than expected or even approached zero [12, 13, 14].

The explanation of this anomalous behaviour is simple but has profound consequences. Basaltic magma at the surface has a density of about 2.7 g/cm^3. When the 1200 °C liquid cools and crystallizes to form feldspars and pyroxenes, the density is about 3.0 g/cm^3. The volume is reduced by about 12% during cooling and crystallization. This contraction is reflected in cracks and cavities in the crystalline product. If a hot, porous-cracked and hence permeable material is placed in a cold liquid (deep ocean water temperatures are normally close to 0 °C) clearly cooling will occur by water circulation [15, 16]. The heat flow patterns near the ridges could obviously be explained by the development of large scale convective cooling cells involving the penetration of cold sea water into hot rock. The thermal gradient in the submarine systems and the permeability are appropriate for massive convective flow of fluid.

Given the idea that such processes must occur, first deep towed temperature measuring devices and later direct submarine observations, found the expected discharge of hot fluids near the submarine volcanic centres. First at the Galapagos spreading centre and now in many ridge situations, hot water has been found discharging at temperatures reaching 300–400 °C. In some cases the venting is cataclysmic reaching temperatures in excess of 400 °C [17]. Because the fluids carry reduced sulphur and metals, metal sulphides (Zn–Fe–S) form during discharge and create the now famous sulphide rich black smoke.

The scale of this water cooling process is impressive. As was shown by Wolery and Sleep [18, 19] and as the most simple calculation of the energetics reveals, about half the energy of the hot basaltic liquid is transmitted to the circulating ocean water. On an annual basis, something like 10^{18} g of ocean water can be heated to 100 °C, or about one third of that mass to 300 °C. Given that the total ocean mass is 1.4×10^{24} g, this means that the entire mass can be processed through the ridges in a few million years, an instant in geologic time.

Before this discovery, most geochemists assumed that the chemistry of the oceans was controlled by the surface runoff from rivers coupled to evaporation–precipitation processes in the ocean reservior. Now a new and significant flux must be considered [19].

The surface runoff of rivers to the oceans is 3.6×10^{19} g/yr [20]. The 100 °C thermal flux is about 10^{18} g/yr. While the latter is smaller, the solubility of many mineral species is so much larger at 100 °C (at times greater than 400 °C), that the deep hydrothermal flux can swamp the cooler surface flux.

Thus studies of ocean ridges have introduced two major new concepts to the surface chemical balance of Earth. First, deep hot water fluxes produce significant additions to the chemical balance of the greatest liquid water reservoir of the planet, the ocean. Second, in the outer shells of the Earth, where rocks are permeable and pores are filled with water, heat may be transported to the surface by fluid flow as well as by conduction. We live on a water cooled planet!

Chemical Separation on and in the Earth

I recently attended a symposium of the Geological Society of America focussed on the topic of the growth of continental crust. A major topic discussed was, that as the mantle forms a surface of basaltic composition, how are silica rich, potassium rich etc. continents produced? The symposium participants were mainly from geophysics. Here I stress that every process on and in Earth, from melting, to hydrothermal cooling, weathering, erosion, biomass concentration, separates chemical components and in a sense purifies. There is no more simple case than that of silica extraction from basalt by hot flowing water to produce chert. For the early hot Earth, the process must have been massive.

Chemical Processes at Ridges:
Mixing of Crust–Hydrosphere–Atmosphere

Ocean water is a moderately concentrated solution of Na^+, Mg^{++}, Ca^{++}, K^+ balanced by Cl^-, SO_4^{--} and HCO_3^-. Most trace metals (Fe, Mn, Pb . . .etc.) are at levels well below parts per million. Sea water is slightly alkaline (in equilibrium with $CaCO_3$) and normally oxidizing with about 4 ppm dissolved oxygen. Silica content is typically near 3 ppm, a level below the input river water because of removal by silica secreting organisms. When sea water flows through hot basalt the processes which occur include:

1. Solution of minerals, both congruent solution (rare in silicates) and incongruent solution or leaching.
2. Mineral precipitation from the solution (e.g. $CaCO_3$ at recharge or SiO_2 at discharge).

3. Oxidation-reduction processes (e.g. rock oxidation by dissolved oxygen or sulfate at recharge, rock reduction by hydrogen at discharge).
4. Ion exchange processes (e.g., albitization) including processes such as $M^+X^- + H^+ \rightarrow HX + M^+$ (including production of halogen acids).
5. Adsorption processes (often the initial step in process 4).
6. Hydration processes.

The basalt produced in the mantle has a low water content, a few tenths of a percent maximum. When ocean floor basalts are sampled from the study of the fragments of ocean floor crust, occasionally found on continents, a large degree of hydration is ubiquitous. At the top of the sea floor column minerals such as zeolites, clays, chlorites are common. At deeper levels minerals of typical greenschist facies (epidote, chlorite, amphibole) dominate. In the deep peridotite level of the ocean crust (3 km +) serpentine is common. Most deep rocks sampled, as in the recent drilling in the Cyprus ophiolite complex, show very high degrees of serpentinization [21]. In general, the water (5–10%) or carbonate content (1–4%) of sea floor rocks increases as one moves off the ridges into regions of older crust and lower heat flow [22].

The fixation of water (and other volatiles like CO_2 and chloride) in rocks of the sea floor is a process of profound consequences for earth dynamics. Almost all ocean floor crust is eventually subducted and returned to the mantle. In this process volatiles are returned to the mantle and such volatiles have a great influence on the mechanics of the mantle and particularly on mantle viscosity. Most models of the development of the atmosphere and hydrosphere discuss early degassing. But at the present time volatile flow must be considered in two directions. If volatiles were not returned to the mantle the volcanic processes above sinking convection cells (subduction zones) might either cease or at least be on a very different scale.

As sea water is almost saturated with $CaCO_3$ and as the solubility of calcite diminishes with increasing temperature, precipitation must occur during the heating cycles. The process $Ca^{2+} + 2HCO_3^- \rightarrow CaCO_3 + H_2CO_3$ will lead to the $CaCO_3$ precipitation commonly observed in altered rocks but at the same time will lead to more acid conditions. Experiments show that when sea water reacts with basalt the pH rapidly moves from 8 to about 5 [13]. The equilibrium CO_2 concentration will eventually be buffered to values appropriate to silicate equilibria:

$$2H_2O + CO_2 + CaAl_2Si_2O_8 \longrightarrow CaCO_3 + Al_2Si_2O_5(OH)_4$$
$$\text{anorthite} \qquad\qquad \text{calcite} \quad \text{kaolin}$$

Thus in general the CO_2 species in sea water will be scrubbed out by carbonate formation. Fresh basalts contain up to about 0.1% CO_2 while typical old altered basalts contain several percent. Again this CO_2 will eventually be returned to the mantle. This mantle volatile recharge process would not occur on a planet without a liquid hydrosphere where surface water cooling was absent!

Free oxygen and sulphate are not in equilibrium with the basalt derived from an essentially reduced mantle. Deep mantle rocks are dominated by ferrous iron and of course, the deepest mantle is in equilibrium with a metallic iron alloy. Observations of the water-cooled lavas of the ocean floor show develop of magnetite (Fe_3O_4) and often hematite (Fe_2O_3). Occasionally, the lavas can even be red where the influx of the cooling water is concentrated. The upper layers of the sea floor invariably show an increase in the Fe^3/Fe^2 ratio. The process involves an oxygen sink, but the quantity of oxygen removed by this process is negligible compared with photosynthetic production (about 5×10^{12} g year^{-1} removed compared to 4×10^{17} g year^{-1} produced). Nevertheless, when 5×10^{12} g is integrated over geologic time and if a higher rate of volcanism is predicted for the ancient earth, the total amount transported is not trivial. The overall process represents a trend via photosynthesis:

$$CO_2 \rightarrow \text{fixed carbon} + O_2$$
$$O_2 + \text{rock} \rightarrow \text{oxidized rock}$$

It is also interesting to note that if photosynthesis was to stop, the atmospheric oxygen mass (about 10^{21} g) would be removed in about a million years given the present rates of fixation.

The other oxidized species present in sea water is also out of equilibrium with basalt. The reaction

$$4H^+ + 11Fe_2SiO_4 \text{ (in basalt)} + 2SO_4^{2-} \rightarrow 7Fe_3O_4 + FeS_2 + 11SiO_2 + 2H_2O$$

has a very large free energy change (-1012.5 kJ) and should lead to almost quantitative removal of sulphate (22). Pyrite and magnetite are ubiquitous in the deeper parts of altered sea floor rocks.

The oceans are close to saturation with barium sulphate and near saturation with $CaSO_4$.

As with $CaCO_3$, the solubility of $CaSO_4$ decreases with increasing temperature. As sea water penetrates hot rock and is warmed, calcium sulphate should be precipitated. Such precipitation has been observed in many samples of such rocks. We have observed massive introduction of sulphate in the pillow lavas of Cyprus. In terms of the global sulphate balance the process is important. The river flux of SO_4^{2-} to the oceans is about 4×10^{14} g year^{-1} and the sulphate removal rate at ridges is similar in magnitude [22].

It is interesting to reflect on this part of the sulphur cycle. Sulphate is introduced into river waters by the solution of evaporites and by the oxidation of sulphides plus a contribution from ocean derived aerosol particles. The sulphate from the oceans removed in the cooling process will form sulphides and be recycled through volcanoes in reduced form. The overall process thus provides another photosynthetic oxygen removal process of considerable significance. In fact, this process alone could remove the atmospheric oxygen reservoir in a few million years. It is interesting to reflect on the possible atmospheric impact of periods of very high basalt production and eruption (see hot spots, below).

The fluids which enter the convective cooling cells are oxidized. Free oxygen and sulphate are rapidly removed. As the fluids become hotter (and we know from the observed discharge temperatures and deep metamorphic patterns and discharge temperatures over 400 °C, that temperatures must reach at least 500 °C in the deepest parts of the system) reactions like:

$$3Fe_2SiO_4 + 2H_2O \rightarrow 2Fe_3O_4 + 3SiO_2 + 2H_2$$

must occur. This process involving the fayalite component of olivine, $(FeMg)_2SiO_4$, is in equilibrium with hydrogen at 1 atmosphere pressure at about 100 °C and as the reaction has a large positive entropy with liquid water, P_{H_2} may attain several atmosphere pressure at depth. The fluids which are discharged will be highly reducing. The hydrogen discharge provides another process which removes free atmospheric oxygen. It is interesting to note that all these processes which could influence atmospheric oxygen have received little attention by geochemists.

River water moves silica into the oceans at an average level of about 10 ppm while ocean water contains silica in the range 3 – 6 ppm [20]. Part of this silica is steadily removed by silica secreting organisms [23, 24].

As sea water penetrates hot rock and is heated, silica must be leached from the rocks. Studies of hot spring discharge, both on continents and in oceans, shows that silica levels tend to be similar to that expected from quartz solubility or even a little higher. The silica content of Galapagos submarine hot springs is around 1300 ppm [13, 25]. While the river flux is much greater than the hydrothermal flux, the much higher silica levels in hot waters shows that this input into the oceans is similar to that of rivers. This is a profound discovery. For this common component of the Earth's crust, the deep flux is of great importance. It was not known two decades ago. It must make us suspect many of the geochemical balances which consider only the surface runoff regimes of our planet.

The concentration of the major cations of sea water, Na^+, Mg^{2+}, Ca^{2+}, K^+ are all strongly modified during the cooling process [22]. As sea water is heated sodium is fixed to form albitic plagioclase ($NaAlSi_3O_8$); potassium is fixed in K-rich clays or even in K-feldspar ($KAlSi_3O_8$) [26]. Magnesium is removed in a complex array of Mg-rich clays and amphiboles and observed discharge water are very low in Mg. While some calcium is removed as carbonate and sulphate at an early stage in the heating process, given that anionic chloride is more or less conserved, electroneutrality is maintained by calcium enrichment in the fluids and in observed Galapagos discharges reaches 4 × sea water values. The end product moves to a calcium dominated chloride solution which is highly reducing. Recent data also shows that for an element like lithium, hydrothermal stripping is highly efficient and the Li flux into the oceans is dominated by the geothermal component [13].

A particularly instructive example involves strontium. Present data shows that the strontium content of the circulating fluid does not change greatly along the circulation path. But exchange of the isotopes $^{87}Sr/^{86}Sr$ does occur. Today, river water carries strontium to the oceans with an 87/86 ratio of about 0.711.

The modern oceans have a ratio of 0.709, while Galapagos discharge has a value of 0.703, similar to the basalts themselves. The reduced ratio in the oceans thus reflects the influence of the hydrothermal discharge [27]. Data from marine carbonates which incorporate Sr, shows that the ocean water ratio has changed significantly over time. Thus in the mid-Jurassic, ocean water had a ratio near 0.707. Such data must reflect the varying ratios of runoff to hydrothermal input, a partial reflection of continental elevations with time [28]. It is interesting to note also, that for sediments of the ancient earth, sea water strontium is essentially that of the mantle [29]. The chemistry of early oceans could have been dominated by geothermal inputs [30].

For the transition metals and many trace metals like zinc, the concentrations carried in thermal waters greatly exceed those in normal surface fluids. Thus for the Galapagos fluids at 350 °C, Fe and Mn reach the 1800 ppm and 1000 ppm levels [13]. Typical river water carry such elements at the 0.1 ppm level or even less. It is clear that for such elements, the hydrothermal flux dominates over the river flux. Dispersion and oxidation must lead to the formation of the Mn-Fe oxide deposits in the oceans which are enriched in elements such as V, Co, Ni, Cu, Zn, Pb, etc. It is also significant that such metals are frequently bio-essential and this flux may have great significance for the health of the marine biomass and influence many biological processes.

Increasingly study of processes at ocean ridges has shown the widespread hydrothermal activity at ocean ridges. Giant thermal plumes have recently been observed over the Juan de Fuca ridge system. As hot fluids discharge, some species like silica will precipitate along the cooling path and quartz vein systems are not uncommon in ophiolites, as observed in samples preserved on land.

As the reduced, calcium and metal-rich fluids approach the surface mixing with fluids at shallow levels must occur. Sulphide formation in pipes and siliceous stockworks is to be expected where Fe–Cu–Zn(Ag) massive sulphide deposits form. But much of the species carried will be dispersed around the discharge vents, where inorganic and biological processes will lead to precipitation in the oxidizing environment.

It has been suggested [31] that massive injection of calcium rich solutions could lead to CO_2 production near ridge vents. The process involves the equilibrium:

$$Ca^{++} + 2HCO_3^- \rightarrow CaCO_3 + H_2CO_3$$

It is also possible that silica injection could have a similar result:

$$4SiO_2 + 3Mg^{++} + 6HCO_3 \rightarrow Mg_3Si_4O_{10}(OH)_2 + 6CO_2 + 2H_2O$$

It is considered that such processes could lead to carbon dioxide production at the rate of 10^{15} g year^{-1}, a rate which would be significant in terms of atmospheric CO_2. Very recently (17) massive plumes have been observed off the Juan de Fuca ridge involving single bursts of 350 °C fluid of 0.1 km^3 volume. Do such processes play a significant part in fluctuations of atmospheric carbon dioxide?

Given that fossil fuel additions (6×10^{15} g year $^{-1}$ C) are perturbing atmospheric CO_2, it must be considered possible that volcanic additions of CO_2 can also be significant.

Ridge-Summary

The processes at ocean ridges, the greatest sites of convective heat-mass transport on the planet, lead to a massive geosphere interaction process. Components from the oceans and atmosphere (Na, K, Mg, S, O_2, U, Cl, H_2O, CO_2 . . .) are added to the solid crust. Components from the mantle-derived basalts (SiO_2, Ca, Mn, Fe, Co, Ni, Zn . . .) are added to the ocean-atmosphere system. Eventually, the modified mantle materials will be returned to the mantle where the global convection cells return to the mantle in the subduction process. Only in the past two decades have we begun to appreciate and quantify these giant geosphere exchange processes.

The Quiet Ocean Floors

Over 70% of our planet's surface is covered by oceans. The near shore continental shelf areas are reasonably well known partly because of the search for hydrocarbon resources. The narrow strips of the ocean ridges are being intensely studied as described above. But the great remaining 60% of our planet's surface is still little known except for the thin blanket of deep sea pelagic sediments. Sedimentation is slow with a major component being wind derived off the continents. Recently, interest has increased in these quiet regions, partly a result of the search for possible nuclear waste sites in the quiet regions covered by thick impermeable mud rocks.

Theoretical studies of heat flow from a spreading centre predict that the high ridge heat floor should decay with time (age) as one moves off the ridge to older rocks according to an age$^{\frac{1}{2}}$ law. As discussed above, the ridge heat flow is greatly perturbed by sea water convection in the hot rocks. But as cooling proceeds, the vigour of convection should rapidly diminish and observed and theoretical heat flow should approach. In a gross way, this behaviour is observed [12]. At an age of about 20 million years, the most dramatic influences of water-cooling appear to decay. But it had been predicted that a way from the new rocks of the ridge to where the basaltic surface is covered by a thin sediment cover, convective circulation might still continue under the less permeable sediment cover [16]. Recent studies of heat flow patterns in older ocean floor [32, 33] have revealed the expected anomalies and it now seems that convective cooling of ocean crust may occur over the entire area of the ocean floors. The significance of this process must be related to the energy available to drive the flow, residual energy from the original hot material of the ridges, plus normal heat flow from the deep mantle.

A problem of global concern involves the thermal gradients necessary to drive fluid convection in a porous medium. The Rayleigh number for such a system is given by

$$Ra = \frac{K\alpha\Delta T\, gH}{K_m^2\, \mu}$$

where K is the permeability, α the coefficient of expansion of the fluid, ΔT the temperature gradient, g the gravitational acceleration, H the thickness of the permeable layer, K_m the thermal diffusivity of the saturated medium and μ is the kinematic viscosity of the fluid [7, 22]. For convection, Ra must exceed a critical value of $4\pi^2$ or 39.48. Clearly the factors which will greatly influence the Rayleigh number are ΔT, H and K. For convection the adiabatic temperature gradient of the system must be exceeded. A careful analysis of the adiabatic gradient in such systems shows that even for gradients as low as $1\,°C\,km^{-1}$, fluid convection is possible [34]. Thus in almost all crustal situations on this planet fluid convection is possible if the other parameters are appropriate. For permeable rocks, even with only normal background mantle heat flow, some degree of deep fluid convection is possible in thick, permeable rock units.

Whenever deep sections of the oceanic crust are sampled by drilling or dredging in fracture zones, the deeper layer of peridotite beneath lavas, dykes and gabbroic intrusives is invariably heavily serpentinized. When obducted ophiolites are studied, high degrees of serpentinization also occur and there is increasing evidence that sea water penetration to depths of the order 10 km is possible, particularly near the ubiquitous fracture zones of the ocean floor. The characteristic serpentinization reaction:

$$2H_2O + MgSiO_3 + Mg_2SiO_4 \rightarrow Mg_3Si_2O_5(OH)_4$$
water + pyroxene + olivine   serpentine

is highly exothermic ($\Delta H = -67\,kJ$). For most such hydration reactions, the ΔH is close to $-41.8\,kJ\,mol^{-1}$ of water. The potential heat generation is not trivial. If $1\,km^3$ of peridotite is serpentinized, the heat generation is in the order of $8.3 \times 10^{17}\,J$ and this process can operate at any temperature within the serpentine field ($500\,°C$ and below) [22]. For comparison, radioactive heat production in granites due to K, U, Th decay is around $8.3 \times 10^{16}\,J$/million years. Thus a low rate of chemical heat production is possible and can be in proportion to the rate at which water penetrates the system. As hydration proceeds a very large volume expansion ($+40\%$) occurs. Such swelling can either seal the system or generate the stress to form new fractures. Study of the fracture systematics of serpentinites and study of water diffusion in serpentine itself, indicates that permeability generation dominates over sealing [35].

Another interesting feature of serpentinization is that the process sea water + peridote can produce as a product, serpentine + salt or more saline solutions [35]. Such highly saline solutions have been found in fluid inclusions at deep levels of the Cyprus ophiolite [36]. It seems unlikely that these were produced by boiling. In general, the solubility of minerals like Fe, Mn, Zn, Co,

Ni in salt solutions increases with salinity and temperature. Warm, very salt solutions provide highly efficient metal leaching systems as observed in the Salton Sea geothermal systems [25]. Such reduced fluids rising beneath a thin impermeable sediment blanket could produce ideal conditions for strata-bound ore precipitation [37].

Recent detailed studies (related to potential for nuclear waste disposal sites) of quiet regions of the ocean floor sedimentary regime, have revealed lateral variation in heat flow, complex faulting patterns and diapiric structures [38]. All could be caused by serpentinization at deeper levels in the oceanic crust. It is obvious that this remaining 60% of our planetary surface deserves increasing attention. It seems possible that the exchange processes characteristic of the ocean ridge environment may continue over the entire system. Perhaps the major differences in this cooler part of the system is that a much greater contribution from solutions more saline than seawater is possible. Such solutions may produce the highly leached white rocks of the sea floor, the rodingites, common in serpentines, where virtually all transition metals and even species like titanium are stripped from the rocks [39].

Subduction: Return Flow to the Mantle

The geochemists of the 1950s considered the rise of volcanic materials from the deep earth, processes which formed the crust, but most models of the geochemical cycle neglected return flow processes. Thus the formation of the outer geospheres (continental crust, hydrosphere, atmosphere) were largely looked on as resulting from a unidirectional process. Over the past decade, new observations from the study of the ocean trenches reveals that such an approach is invalid.

New ocean crust is formed at a rate of about $15 \text{ km}^3 \text{ year}^{-1}$ at the ocean ridges. This mass of basaltic material ($4.2 \times 10^{16} \text{ g y}^{-1}$) can be compared with the total mass of the Earth's crust $2.6 \times 10^{25} \text{ g}$ from which it is apparent that the present rate of crust production by ridge magmatism alone would form the total crust in 10% of Earth history. Given that the rate of magma generation must have been much faster for the hotter early planet it is clear that the crust, particularly the oceanic crust, has been recycled many times.

There are very few fragments of old oceanic crust present on Earth. Oceanic crust is recycled at a rate essentially equal to its production rate. This fact has at times been called one of the laws of plate tectonics. The recycling process, sinking of convection cells, subduction, is associated with the great topographic lows of the ocean trenches which are also normally lows in global heat flow.

Seismic studies of the oceanic lithosphere and earthquake generation in trench regimes shows that the oceanic lithosphere, about 100 km thick, bends and sinks or is thrust into the mantle and that the process is essentially initiated along the ocean trench axes. The old ocean lithosphere is cold and heavy and to sink must be denser than the surrounding upper mantle. Angles of this underthrusting process are highly variable from almost vertical to only a few degrees [40]. The exact balance of forces which causes and initiates subduction is still not well resolved. Further the depth to which sinking occurs is still a matter of debate but sinking to at least 700 km, where a major density change occurs in the mantle, is generally accepted.

For the geochemist the key problem regarding the subduction process is to quantify exactly what is subducted. We can be sure that the igneous material of old ocean floor is more or less quantitatively subducted. We know that the sea floor cooling processes at ridges, changes the chemistry of what originally came from the mantle in important ways. The near ridge processes are rather well known but much remains to be documented before quantification is adequate. The later, slow, alteration processes on the quiet ocean floor conveyer belt, are not well documented.

But we can be sure that subduction must recharge the upper mantle with volatiles like H_2O, CO_2, S species, N species, halogens, and the subducted material is also slightly more oxidized. Today, most would agree that the mantle must be recharged with components like Na, K, Mg, and trace elements like uranium. In summary, components of the continental crust transported to the oceans, and atmospheric components, tend to be mixed into the sea floor rocks pre-subduction. There are very important gaps in our knowledge. We need to know how much serpentine is subducted, for altered peridotite is a massive H_2O carrier (as it may be for CO_2). But modern ocean floor studies, and data from continental ophiolites, all show massive serpentinization [41, 42, 43].

For decades there has been great argument concerning the fate of sediments in the subduction process. Gilluly [44] (and many other trench volcanologists) suggested that there simply was not enough accumulated sediment in trench regimes and that some must be dragged into the mantle along with the sea floor rocks. Other have stressed the lack of typical aluminous rocks and pelagic sediments which might be scraped off in trench regimes. On the other hand, geochemists and isotope geochemists, had difficulty in finding convincing evidence for massive subduction of sediments. Careful consideration of the density balance have shown that major sediment subduction, or other light materials, is possible [45].

The geochemical problem is tricky. If the mantle chemical reservoir is unique (and similar to some classes of meteorites, etc.) the problem would be simple. But if continuous mixing of crust-hydrosphere and mantle occurs, the mantle, and particularly its most fusible parts, will represent the result of steady state mixing. The mixing process will tend to reduce in intensity with time as the power sources decay in the interior. New high resolution seismic data on the upper mantle increasingly shows complexity in the upper mantle as would be expected from left over unmixed subduction fragments [46, 47, 48].

Many of these problems are being resolved with increasing observations in trench regimes by seismic studies, the new side scan sonar studies of topography, and increasingly direct observation from submersibles and drilling [49]. The new observations show that as the lithosphere bends during its descent, it cracks, forms major horst and graben structures with horst elevations at times more than 1 km [40, 49]. Pelagic sediments have been located by seismic methods and found on horst tops, graben floors, and even under the continental edge. The massive lithosphere cracking in the subduction regimes has provided the mechanism of mechanically locking light sediments into the descending lithosphere. The recent studies of the Japan trench by all methods, including submersibles, has demonstrated that most of the deep Pacific sediments are subducted [50, 51]. There is essentially no accretion of sediment on the edge of the Japanese continent. Thus K rich, H_2O rich, even U-rich, materials can be returned towards the mantle. In some subduction regimes with low angle subduction, most sediment is accreted [52].

It seems that the quantity of sediment likely to be subducted will be related to the roughness of the bending lithosphere as it is subducted. This will be influenced by the angle of subduction and the subduction velocity, and the age of the subducted lithosphere. Further, the amount of sediment accretion must also depend on the local sediment thickness and local sedimentation rate on the sea floor.

That sediment is removed by subduction is slowly becoming accepted. Figures like $1-3$ km^3 year^{-1} $(2-6 \times 10^{15}$ g year$^{-1})$ are considered possible [53]. Given that the mass of continental crust is about 1.6×10^{25} g, and a sediment subduction rate of 6×10^{15} g year^{-1}, the continental crust will be recycled through the mantle in about 3 billion years. This is quite slow but given faster recycling rates in a more energetic young Earth, the concept of the permanency of light continental material, unidirectional growth of continents must be questioned.

However, there is no longer any doubt that a degree of continental crust, hydrosphere, recycling must now be built into models of the geochemical dynamics of the Earth. The processes are not well quantified and probably will not be until we have better sampling of the ocean floors and the trench regimes of the world. Quite recently [54] studies of the fracture zones in parts of the Indian Ocean have surprised people. These studies reveal fractures several hundred km long, 30 km wide, and up to 6 km deep in the Indian Ocean floor (even greater than the Grand Canyon!). Such a fracture must close or be sediment filled pre-subduction. It could carry a sediment wedge of 36,000 km^3 volume into the mantle. It is no surprise that as seismologists probe the upper mantle in greater detail that local heterogeneities are revealed [47]. And it should be stressed that the ocean crust contains a myriad of fracture zones.

As oceanic crust, with varying amounts of sediment, moves into the mantle it is heated and compressed. Volatiles will be expelled at several stages. Early in the process pore fluids will hydrofracture their way to the surface lubricating the subduction thrust planes. Oil-gas etc. in the sediments will also be expelled early in the process. We have observed hydrocarbons coating fault planes above the subduction zone on Vancouver Island, Canada, where the Juan de Fuca plate is being subducted. Exotic fauna similar to that at ridges has now been found in deep trenches [51].

Water and volatiles contained in minerals (clays, chlorites, epidotes, serpentine, amphiboles, carbonates, etc.) will be expelled as the rocks pass through the appropriate metamorphic isograds for a low T high P path (zeolite facies → blue schist facies → eclogite etc.) [55, 56]. As normal for any prograde metamorphic series, fluids will be evolved over a P-T range except that for some simple reactions involving phases like serpentine, almost univariant bursts of fluid will occur. The fluids can migrate by hydrofracture mechanisms into the overlying plate or along the thrusts. The movement of the high-P fluids can lead to extensive metasomatic processes with a large array of vein minerals characteristic of blue schist eclogite terraines.

Dehydration and fluid transport back to the hydrosphere must be essentially quantitative or there would be evidence for declining ocean volumes (the present water subduction rate in minerals is about $1 \text{ km}^3 \text{ year}^{-1}$. As the ocean mass is 1.4×10^{24} g, these processes will recycle the hydrosphere in about 1 billion years). depends critically on the amount of serpentine (15% H_2O) subducted and this is not quantified at the present time.

An important process that must occur with the K-rich portion of the upper surface of the lithosphere involves the reaction with the warm ultramafic mantle above the subduction surface which can produce minerals like phlogopite, $KMg_3Si_4O_{10}(OH,F)_2$.

The overlying mantle at great depth is hot, and the injection of water into the overlying plate must lower the viscosity and promote mantle convection above the subduction thrusts [57]. This process now leads to another important heat flow process catalyzed by fluid, volatile, injection. The entire process has been beautifully imaged by recent deep electric sounding showing high conductivity zones along the thrust planes and in the overlying mantle [58].

Volatile-fluxed rising mantle convection cells can melt and eventually produce the characteristic product of subduction volcanism, andesite. This melt is in a general way similar to the basalt of the ocean ridges except that it tends to contain more water, silica, potassium, uranium..., exactly what would be expected for mantle contaminated by a solute rich high pressure fluid from the ocean floor complex. In general too, the $^{87}Sr/^{86}Sr$ ratio is higher, more crustal, than for a ridge basalt as in the $^{18}O/^{16}O$ ratio [59, 60, 61].

Recent studies have shown that the interesting species ^{10}Be can also be enriched. This short-lived beryllium isotope is synthesized by cosmic ray processes in the atmosphere. The ^{10}Be is implanted in the surface of marine sediments. Thus its presence in andesite volcanoes provides convincing evidence for the rapid deep subduction of the skin of the ocean floor crust. Interesting results have also been found for boron [62].

Very frequently, as in Pacific Rim trench systems, the subduction zone volcanism occurs near the edge of continents and under continental crust. The rising hot mantle produces andesite volcanism and melting of the base of the crust. The product of this process are the plutonic-volcanic series of the granite-granodiorite-tonalite groups; all high silica viscous magmas with significant water contents.

A recent result from experimental petrology of profound importance has involved the study of the compressibility of magmas. Naturally, magmas are more compressible than their crystalline equivalents [63]. Basaltic magmas at typical continental Moho pressures (near 10 kbar) can be more dense than average continental crust. Thus such magmas may float off the continental crust near the Moho region and underplate the crust with basaltic (gabbroic) materials. Obviously the emplacement of 1200 °C magmas beneath continental crust will enormously perturb the normal gradient of about 20 °C km^{-1} (for a crust 30 km thick, the Moho temperature is normally about 500 °C).

Underplating can lead to a number of very significant mixing and mass transport events [64]. Dense crustal materials may founder into the underplate magmas, be assimilated or melted and lead to mixed magma products. As such materials will normally be in the amphibolite facies with a significant H_2O-CO_2 content, very high temperature fluids may be produced and injected into the overlying crust. As massive carbonates are more resistant to change than hydrated systems (H_2O a better reaction catalyst than CO_2) one would expect that such fluids would be at first H_2O-rich and that later fluids would be dominated by CO_2. This has been observed in the deep fluids of very high-T metamorphic rocks where almost pure CO_2 is found in fluid inclusions.

Given a Moho T strongly increased by mafic underplate, continental crust will melt producing granite-granodiorite-tonalite melts depending on the local composition. Such melts may mix with underplate magmas to varying degrees depending on the nature of the melting zone, but in general a high and low silica magma series is produced. This produces the characteristic suites of deep granitic plutons and erupted rhyolites and deep gabbroic plutons (and anorthosites) and erupted basaltic andesites. The detailed compositions are highly variable exactly as would be expected for such a process.

Above the zone of magma underplating and crustal melting a vast prograde metamorphic process must follow as the entire crust responds to increased heat flow. The volatiles in large sections of crust will be expelled upwards and create the great vein swarms characteristic of such terraines. As magmas rise to high levels, they may create fluid cooling systems in the water-saturated, porous cracked zones of the upper crust. Such systems are quite analogous to the ocean ridge processes except that normally the fluids are less saline (there are very important exceptions such as the Salton Sea system of S. California and it should be stressed that evaporites are very common in high level sedimentary series). Again a very important fluid transport process by high-T fluids can influence the surface chemistry of regimes near the volcanic belts. Compared to the ridge systems, one would expect the overall global impact to be about 10% of the ridge fluxes or in relation to the overall magma volumes. But as acidic plutons can be an order of magnitude larger than typical ridge magma chambers, local influences may be extreme.

Finally, it should be noted that the fluid transport processes associated with the continental volcanic events lead to very diverse types of geochemical enrichments or ore-forming processes (As, Sb, S, W, Sn, Cu, An, Ag, Fe, Mn, U, Ta, ...). Complex magma fluid systems are being increasingly recognized [65].

Continental Collisions and Global Geochemistry

The most common process which balances the new crust production at ridges is the sinking, subduction, underthrusting of oceanic lithosphere at the sites of the great ocean trenches. But at times, the ocean conveyer belt carries a continent with it and eventually, masses of continental crust and lithosphere may collide. The process is not infrequent in the geologic record and one of the greatest events is currently occuring as continental India has collided with Tibet. But this collision zone is only a part of the entire Alpine system that crosses Southern Europe and extends into Northern India.

Molnar and Gray [45] examined the question as to whether or not a continental edge could be subducted along with oceanic lithosphere. Their conclusion was that density relations could allow this if the continental crust was strongly coupled to the lithosphere. In the Himalayan example, the crust has been thickened through a series of giant thrust slices [66–67]. In the Tibet-Nepal region thickened crust, 60–80 km thick, has been created over an area of 1500×3000 km (60% of the area of continental Australia). The process still continues as India moves north at about 5 cm year^{-1}. The global influence of such an event is considerable. The creation of the high topography of the Himalayas influences global weather patterns and some consider that the present monsoon weather patterns was initiated by this event.

When two continental blocks are stacked to double thickness major fluid transport processes must result. When a major continental block moves over or under another a series of fluid transport processes must follow.

Very rapidly, fluids will be squeezed out of the top of the underthrust block. Oliver of Cornell terms this the squeegie effect [68]. The dominant fluid will be water-rich but if salt basins or hydrocarbon basins are present (and they are not uncommon) oil and gas, or salt fluids may be squeezed outwards and upwards along the thrusts on a vast scale, a process which provides the fluid pressure to catalyze the thrust motions.

For the scale of the Himalayas, the underplate has an area of 4.5×10^6 km^2. Given that the upper crustal rocks may contain 10% H_2O, the fluid volume squeezed out can be of the order of 2×10^6 km^3 (2×10^{21} g) which equals 0.1% of the ocean volume or about half the mass of the present ice caps (3×10^{21} g). There will obviously be a large infuence on global sea level, even a potential influence on ocean salinity. Because the fluids are coming from upper crustal levels, they will have higher $^{87}Sr/^{86}Sr$ ratios than ocean water and must contribute to changes of this ratio in the oceans. For a thrust of Himalayan scale moving at 10 cm year^{-1}, fluid expulsion could attain 0.5 km^3 year^{-1}.

Given that the normal thermal gradient in continental crust, which is enriched in U–Th–K, is about 20 °C km^{-1}, the entire thickened crust (60 km $\pm$) must begin to warm up and temperatures near the base should rise towards 1000 °C. In the underthrust regions, a vast prograde metamorphic event must occur influencing the entire 30 km underthrust section. This metamorphic event will again produce a fluid event with volumes again approaching present ice cap volume. The fluids evolved in the temperature range 100–600 °C will transport high concentrations of solutes and should again lead to massive production of geochemical anomalies in the overthrust rocks. Near the thrust region where the thermal gradient is initially inverted, some fluid will be absorbed in retrograde metamorphic processes in the overthrust crust [69, 70, 76].

Finally, at deep levels in the thickened crust, melting must occur producing "pure" crustal granites and perhaps causing the large conductivity anomalies. These are widely observed in the present Himalayas and the young granites have a chemistry (with extreme $^{87}Sr/^{86}Sr$) and isotopic systematics as would be expected for water rich crustal melts [72, 73, 74, 75].

The same types of "squeegie" effects can be observed on smaller scales at "conservative", plate boundaries where major strike-slip fault motions occur. Typical is the great Alpine fault of New Zealand where override leads to crustal thicknesses of Himalayan magnitude [64, 76]. Again, fluids expelled are tightly linked to the dynamics of motion.

Hot Spots

While most global melt production occurs at ridges, subduction and collision zones, hot spot or random volcanism is locally of great significance. Such volcanism, normally basaltic, produces something like $1\,km^3$ of magma annually.

In a sense, hot spots reflect random thermal noise to the large convection structures of plate tectonics [77]. Inspection of any global map, particularly that for the ocean basins, indicates that the areas of crust influenced by hot spots is non-trivial, well over one hundred occur across the globe. As shown in Turcotte and Oxburgh [77], a deep mantle plume leading to a hot spot may influence the thermal structure of the lithosphere over areas of the order of $100,000\,km^2$. If this is multiplied by the number (say 100) the global area influenced approaches 10 million km^2, 7% of the surface. The impact of hot spots in fluid circulation is quite unexplained and unquantified.

The ultimate source of "hot spot" magmas is generally assumed to be at much deeper levels in the mantle than for the common ridge basalts [3]. Periods of massive eruption are highly irregular [78, 79]. For example, the famous Deccan floods basalts of India which covered an area of at least $500,000\,km^2$, were erupted over a period of 2 million years at an average rate of $1\,km^3\,year^{-1}$. Some consider that the correlation with magnetic reversals may indicate a connection with processes near the core-mantle boundary and some believe that these processes may be related to periods of massive extinctions or "gradual volcanic catastrophes".

Outpourings of massive quantities of basalt (or other volcanics) on continents may have very significant effects on the global environment. First they may generate massive cooling systems over wide areas which can influence H_2, HF, HCl . . . , injection into the atmosphere. They also place vast quantities of reduced materials on the Earth's surface. Most volcanics are rich in Fe^2 which is subject to oxidation to form Fe^3 oxides. An intriguing question is as to whether or not there can be significant changes in atmospheric oxygen (or even ozone!) resulting from such events (for example, 1 km^3 of basalt, 3×10^{15} g, containing 0.3×10^{15} g of FeO, could absorb 3×10^{13} g of oxygen to form Fe_2O_3. The oxygen content of the present atmosphere is 10^{21} g. The photosynthetic oxygen flux into the atmosphere is about 4×10^{17} g year^{-1}).

Mixing and Tectonics

In recent years studies of the detailed geochemistry of the accessible mantle products, solid inclusions and melts, have been used to estimate the degree of crust-mantle interactions in all major parts of the plate tectonic process. The trends in the conclusions of all such studies is to indicate that such interactions do occur but to highly variable degrees. Studies of specific systems like samarium-neodymium [53], stable isotopes like O–S–C [80], helium isotopes [81], rare earth elements [82, 83] have been of particular importance.

It is interesting to note that among the primoidial elements, perhaps coming from the lower mantle or core [84], only helium and its ^{3}He isotope appear to represent original materials from the planetary formation process. But even in this case there is reason to question the source. It has been found [85] that ^{3}He is concentrated in deep-sea sediments, a result of implantation from external sources (cf. ^{10}Be). Present analysis shows that subduction of such sediments can hardly account for the quantities needed for deep source ^{3}He. But one must be cautious. If the ^{3}He source is external it will also be highly irregular in mass and time. It will be interesting to observe the evolving story of such extraterrestrial species in future years.

The Creation of High Topography

The convective transport of energy and mass towards the surface is related to the formation of Earth's topography. It is interesting to reflect on what would be the state of our planet and its biosphere if such processes were to cease, if cooling was by conduction only. Above, we have considered the very significant fluid transport fluxes associated with magmatic cooling and metamorphic events. The creation of high topography on continents associated with subduction, collision and hot-spot events must lead to very large fluxes driven by gravitational flow in the porous and fractured rocks associated with high topography, the difference between precipitation, evaporation, and surface runoff [65].

At this time there is little data concerning deep fluid flow on continents but studies, often related to oil-gas reservoirs, or from deep nuclear waste disposal studies are showing that such processes may be of greater importance than previously considered. Just as for ocean ridges, studies of heat flow with hydrologic parameters increasingly show the influence of deep flow cells in the evolution of water systems, with penetration to depths such as 15 km [85]. An intriguing problem involves the estimation of how much fluid finds its way to oceans by paths other than rivers.

Atmosphere-Hydrosphere-Crust-Biosphere Interactions

The processes which involve interactions between liquid water, the atmosphere, the solid surface (and the porous, permeable, outer surface layers) and the living cells of the biosphere have a profound influence on all parts of geochemistry [23, 86]. The roles of each part of this system (now called by some the ecosphere) are tightly linked and understanding of these linkages is the focus of the great new scientific programme of the International Council of Scientific Unions (ICSU) termed the International Geosphere Biosphere Project [19].

In terms of the topic of this essay, geosphere interactions:

The Atmosphere moderates the radiation field (and in part, with our magnetic field, the particle flux of cosmic rays and solar wind) from the Sun. It provides the Earth's thermal blanket which maintains the liquid water on the surface, and supplies the transport system of CO_2–O_2–H_2O and dust to all parts of the Earth's surface. It is the great reservoir of free oxygen, a result of biological processes.

The Hydrosphere is an excellent solvent for most minerals. It transports materials in dissolved and particulate form to the great reservoirs of lakes and oceans and provides the growth medium for the bulk of living species which require a liquid dominated by water inside and outside their cells. Reactions of liquid water with surface materials modifies the chemical composition by congruent or incongruent leaching processes and most surface water is saturated with respect to atmospheric O_2 and CO_2. The hydrosphere reservoirs are also the great sites of the formation of new minerals, often hydrated, and oxidized, which are in equilibrium with the surface environment [87]. The mineral formation is very frequently bio-catalyzed.

The Biosphere in large part controls atmospheric oxygen and moderates atmospheric carbon dioxide via photosynthetic processes. Massive transport of inorganic components occurs via the biosphere which is also associated with wide ranging types of biomineralization processes [23, 83, 84, 86, 87, 88, 89]. We know that organisms can live in liquids (salt) below 0 °C and up to temperatures a little above 100 °C. There is increasing evidence that organisms are present in deep ground water systems but little is yet known about such systems. The carbon in biota is estimated to be about 560×10^{15} g (something like 1000 km³ of living cells) which contain a host of major and minor essential and non-essential inorganic components. As the cycling rate of these cells is something like 10^{17} g year^{-1} and given 1% inorganics, 10^{15} g of rock components are cycled annually through cells. This is a massive process. We know that life has existed throughout the recorded history of the planet, 4000 Ma, and if this cycling rate 10^{17} g year^{-1} is assumed constant, the integrated mass of living cells $(4 \times 10^{26}$ g) approaches the mass of the Earth $(6 \times 10^{27}$ g).

Soil which forms on all continental surfaces is the complex product formed by the action of biota, water and air on surface rocks. It is normally a highly porous medium, containing living and dead organic matter, liquid water and air and primary and secondary minerals which are dominated by clays and metal oxides. Studies of weathering of new volcanic rocks shows that time to produce a medium suitable for sustained agriculture is of the order of 10,000 years [90].

Atmosphere-Surface transport involves the great exchange processes of the carbon-water cycles and the oxygen cycle which is tightly coupled to the carbon cycle. Also of great biological interest are the nitrogen-phosphorous cycles [91]. The classical views on these cycles involve the massive exchange processes which influence the biomass and normally much less attention is devoted to processes which involve the solid earth. It will be noted for the carbon cycle that the present increase in the atmospheric component (0.4% per year) is normally attributed to fossil fuel combustion which makes up only 3% of the flux into the atmosphere. This clearly shows the sensitivity of the total balance. One of the future challenges of geochemistry is to quantify the mass and fluctuation of the inputs and outputs from the solid earth. The same comment can be made for all the cycles. It has been noted early that periods of intense volcanism followed by weathering could even perturb the oxygen cycle.

Another fascinating part of the geochemical cycle involves wind transport of materials a subject of increasing interest as soil erosion gains more attention as man invades increasingly difficult or less desirable terrains for agriculture.

Current estimates show that about 5×10^{14} g of dust of small particle size (less than 20 µm) is transported annually [92]. Dust and aerosols play a very significant role in atmospheric properties (light reflection, provision of catalytic surfaces, etc.). It is estimated that atmospheric aerosol production is about 2.5×10^{15} g year^{-1}, which is dominated by ocean salt and about 20% mineral debris. Given that almost 1 km^3 of ocean salt is so distributed annually, it is clear that this has geochemical significance. Periodic volcanic eruptions can greatly enhance dust injection. It has recently been found that in some heavily weathered lateritic terrains certain geochemical parameters can be limited to inputs from dust. Finally, it is well known that dust makes up a significant contribution to pelagic sediments. Dust inputs from the Sahara into the N equatorial Atlantic have been placed at up to 4×10^{13} g year^{-1}.

As noted above, the living biomass, about 1000 km^3 in volume can play a very significant role in surface chemistry. About one third of the elements in the periodic table have been shown to be bio-essential [93] and this number tends to increase as the science of molecular biology develops. Such elements can accumulate in organic debris (oil-coal-peat, etc.) or be released following transport of the bio-materials. Typical are elements such as Cl, Br, I, B, S, N, P, Fe, Zn, Mo, Cn, V, Ni, Mn, Cd, Ag, Pb, U . . . In addition biomineralization processes can play a dominant role in sedimentary mineral formation (carbonates, phosphates, oxalates, silica, Fe-Mn oxides, fluorides, sulphides, clays, sulphates, etc.).

Continents are eroded by wind and water transport. The present river particulate flux is placed in the range 15×10^{15} g year^{-1}. The load of dissolved solids which averages about 100 ppm, given a global runoff of 3.7×10^4 km^3 year^{-1} (3.7×10^{19} g year^{-1}), transports another 3.7×10^{15} g year^{-1} for a total of 18.7×10^{15} g year^{-1} or approximately 7 km^3 of continental crust is dissolved and fragmented and transported per year [87]. Given that the mass of the crust is 2.6×10^{25} g and about half is continental crust, this represents reworking of the continental mass about every 500 million years. This simple observation calls our attention to the efficiency of metamorphic processes associated with the convective forces which sweep up the sedimentary debris to produce new crystalline metamorphic and igneous rocks.

Most of the primary sediments are water-carbonate (carbon) rich. The sediment forming rates and metamorphic-melting reworking rates must be comparable and point the direction for estimation of the deeper fluxes into the atmosphere-hydrosphere system.

Deep Recycling

As discussed above, the present subduction process provides a mechanism for return of significant quantities of surface materials, more or less in equilibrium with the atmosphere-hydrosphere-biosphere systems, into the deep earth. The deep recycling rates are important in terms of upper mantle chemistry and its convective properties. The recognition of this process leads to interesting questions regarding the balance of H_2O–CO_2 and other volatiles on our planet. For example, at present the deep water recycling rate is something like 1 km^3 year^{-1} which will process the ocean mass in about 1 billion years [94, 95]. Other processes related to crustal delamination in the deep parts of collision zones, hot spots and sites of magma underplating, are little understood but crustal reworking and partial subduction is a major process with almost a third of the present continental crust being reprocessed today. The recent finding of very high pressure phases in metamorphic rocks, like coesite [96] and diamond [97] clearly show how deep crustal materials can be dragged and how rapidly returned to the surface. It should be noted that there may be problem with very small nanometer-sized diamonds which, because of surface energy influences, may be more stable than graphite [98].

In a general way, degassing an object (a positive entropy process) is a response to increasing temperature. But if the Earth is cooling, and being mixed, there must be a tendency for the mantle to hold increasing quantities of volatiles in solid phases. Have the volumes of oceans remained constant or have they changed our Earth history? The same questions can be asked about the mass of continental crust over time. This high silica material is not in equilibrium with the mantle. The balance depends on the extrusion of the low melting fraction of the mantle and return flow via subduction. Such questions can probably be approached using the modern power of geochemical and isotopic techniques.

References

1. Mason B (1966) Principles of geochemistry, John Wiley, New York, p 329
2. Anderson DL (1989) Theory of the earth, Blackwell p 366
3. Dziewonski AM, Woodhouse JH (1987) Science 236: 37
4. Gough DI (1981) Am Geophys Union Geodynamic Series, 5: 87
5. Bowring SA et al. (1989) Geology 17:971
6. Nutman AP et al. (1989) Canadian J Earth Sciences, 2159
7. Turcotte DL, Schubert G (1982) Geodynamics: Applications of continuum physics to geological problems, John Wiley, New York, p 450
8. Broecker WS (1985) How to build a habitable planet, Eldigio, New York, p 291
9. Mutter JC et al. (1988) Nature 336: 156
10. Burnett MS et al. (1989) Nature 339: 206
11. Macdonald KC et al. (1989) Geology 17: 212
12. Emiliani C (1981) The oceanic lithosphere, John Wiley, New York, p 1738
13. Rona P et al. (1983) Hydrothermal processes at sea floor spreading centres, Plenum p 796
14. Ribando MR et al. (1976) J Geophys Res, 81: 3007
15. Lister CRB (1974) Geophys J Rog Ast Soc, 39: 465
16. Lister CRB (1977) Tectonophysics 37: 203
17. Baker ET et al. (1987) Nature 329: 149
18. Wolery TJ, Sleep NH (1976) J Geol 84: 249
19. Fyfe WS (1988) Bull Geol Inst Uppsala 14: 13
20. Berner EK, Berner RA (1987) The global water cycle, Prentice-Hall, New Jersey, p 397
21. Gibson IL et al. (1989) Geol Survey of Canada, Paper 88–9, p 393
22. Fyfe WS Lonsdale P (1981) in (12), 589
23. Degens ET (1989) Perspectives on biogeochemistry, Springer, Berlin Heidelberg New York, p 415
24. Ferris FG et al. (1986) Nature 320: 609
25. Fyfe WS et al. (1978) Fluids in the Earth's crust, Elsevier, Amsterdam, p 383
26. Munha J et al. (1980) Contrib Mineral Petrol 75: 15
27. Barrett TJ, Jambor JL (1988) J Min Soc Canada 26: 429
28. Burke WH et al. (1982) Geology 10: 516
29. Veizer J, Jansen SL (1979) J Geol 87: 341
30. Fyfe WS (1980) Chem Geol 30: 341
31. Owen RM, Rae DK (1985) Science 227: 166
32. Davis EE, Lister CRB (1977) J Geophys Res 82: 4845
33. Anderson RN et al. (1979) Science 204: 828
34. Straus JM, Schubert G (1977) J Geophys Res 82: 325
35. Macdonald AH, Fyfe WS (1985) Tectonophysics 116: 123
36. Vibetti NJ et al. (1989) Geol Surv Canada, Paper 88-9: 229
37. Barriga FGAS, Fyfe WS (1988) Chem Geol 69: 331
38. Shephard CE et al. (1988) Sandia Report, SAND87-1913. UC-70, p 303
39. Barriga FGAS, Fyfe WS (1984) Contrib Mineral Petrol 84: 146
40. Uyeda S (1983) Episodes 1983: 19
41. Moores EM, Vine FJ (1988) Geol Soc Am Bull 100: 1205
42. Fryer P et al. (1985) Geology 13: 774
43. Auzende JM et al. (1989) Nature 337: 726
44. Gilluly J (1971) Geol Soc Am Bull 82: 2387
45. Molner P, Gray D (1979) Geology 7: 58
46. Davies GF (1979) Earth Planet Sci Letters 44: 231
47. Shearer PM (1990) Nature, 344: 121
48. Wood B, Helffrich G (1990) Nature 344: 106

49. Hilde TWC, Uyeda S (1983) Tectonophysics 99: 85
50. Lallemont S et al. (1986) La Recherche 182: 1344
51. Le Pichon X (1986) Kaiko, Voyage aux extremites de la mer, Odile Jacob, Paris, p 246
52. Davis EE, Hyndman RD (1989) Geol Soc Am Bull 101: 1465
53. De Paolo DJ (1988) Neodymium isotope geochemistry. Springer, Berlin Heidelberg, New York
    p 187
54. Emmerman R (1989) Personal communication
55. Fyfe WS, McBirney AR (1975) Am J Sci 275A: 285
56. Lewis TJ et al. (1988) J Geophys Res, 93: 207
57. Silver PG, Carlson RW (1988) Am Rev Earth Planet Sci 16: 477
58. EMSLAB group (1988) Eos , 69 No. 7, Feb. 16: 89
59. Thorpe RS (ed.) (1982) Andesites orogenic and related rocks, John Wiley, p 724
60. Norman MD, Leeman WP (1990) Chem Geol 81: 167
61. Kenji N et al. (1989) Geochemical Journal 23: 45
62. Morris JD et al. (1990) Nature 344: 31
63. Herzberg CT et al. (1984) Contrib Mineral Petrol 84: 1
64. Fyfe WS (1987) Ore geology reviews 2: 21
65. Torgersen T (1990) Eos, 71: No. 1, 1
66. Mascle G et al. (1990) La Recherche, 21: No. 217, 30
67. Allegre CJ et al. (1986) Nature 14: 99
68. Oliver J (1986) Geology 14: 99
69. Fyfe WS (1981) Geodynamics Series, Am Geophys Union 5: 82
70. Fyfe WS, Kerrich R (1985) Chem Geol 49: 353
71. Fyfe WS (1985) Tectonophysics 119: 29
72. Ferrara G et al. (1983) Geol Rundsch 72: 119
73. Pham VN et al. (1986) Nature 319: 310
74. Fyfe WS (1988) Trans R Soc Edinburgh 79: 339
75. France-Lanord C, LeFort P (1988) Trans R Soc Edinburgh 79: 183
76. Allis RG (1981) Geology 9: 303
77. Turcotte DL, Oxburgh ER (1978) Phil Trans R Soc London 288: 561
78. Rampino MR, Stothers RB (1988) Science 241: 663
79. Courtillot V et al. (1988) Nature 333: 843
80. Faure G (1986) Principles of isotope geology (2nd edn), John Wiley, p 589
81. Mamyrin BA, Tolstikhin IN (1984) Helium isotopes in nature, Elsevier, p 273
82. Henderson P (ed.) (1984) Rare earth element geochemistry, Elsevier, p 510
83. Lipin BR, McKay GA (eds) (1989) Geochemistry and mineralogy of rare earth elements, Min
    Soc Am, Reviews in Mineralogy 21: p 348
84. Fyfe WS (1980) Geol Assn Canada, Special Paper 20: 77
85. Nesbitt BE, Muehlenbachs K (1989) Science 245: 733
86. Ittekkot V et al. (eds) (1990) Facets of modern biogeochemistry, Springer, Berlin Heidelberg
    New York, p 433
87. Berner EK, Berner RA (1987) The global water cycle, Prentice-Hall, New Jersey p 397
88. Nancollas GH (ed.) (1982) Biological mineralization and demineralization, Springer, Berlin
    Heidelberg New York, p 415
89. Leadbeater BSC, Riding R (eds) (1986) Biomineralization in lower plants and animals, Systematics Association, Clarendon, Oxford, p 401
90. Colman SM, Dethier (eds) (1986) Rates of chemical weathering of rocks and minerals, Academic,
    p 603
91. National Research Council, U.S.A. (1986) Global change in the geosphere-biosphere, National
    Academy, Washington, D.C. p 91
92. Greeley R, Iversen JD (1987) Wind as a geological process on Earth, Mars, Venus and Titan,
    Cambridge University, p 333
93. Mertz W (1981) Science 214: 1332
94. Ozima M (1987) Geohistory: global evolution of the Earth, Springer, Berlin Heidelberg New
    York, p 165
95. Fyfe WS (1983) Tectonophysics 99: 271
96. Chopin C (1984) Contrib Mineral Petrol 86: 107
97. Sobolev NV, Shatsky VS (1990) Nature 343: 742
98. Badziag P et al. (1990) Nature 343: 244

Environmental Inorganic Geochemistry of the Continental Crust

Harald Puchelt
Institut für Petrographie u. Geochemie der Universität
7500 Karlsruhe, FRG

Introduction. 29
Natural Concentrations of Elements in the Continental Crust 29
Magmatic, Metamorphic, and Sedimentary Rocks 33
Radioactive Elements. 34
Soils as Links Between the Lithosphere and Biosphere. 35
Systematics of the Pedosphere. 37
Use of the Pedosphere in Geochemical Exploration for Deposits 37
Geochemical Distribution of Elements in Soil Horizons
   and Catena. 38
Upper Limits for Concentrations of Elements in Soils
   for Agricultural Use. 43
Anthropogenic Contamination of Soils with Inorganic Elements
   and Their Distribution in Soil Horizons 45
Emission from Industrial Sources 45
Mining Industries and the Connected Metallurgical Industry 47
Coal-Fired Power Plants. 50
Cement Factories. 50
Fluorine-emitting Industries. 53
Mercury-Emitting Industries 53
Contamination by Fertilizers (Mineral Fertilizers, Sewage Sludge,
   and Compost). 54
Contamination by Pest Control Agents (Herbicides, Fungicides,
   Insecticides) 55
Contamination by Atomic Bomb Tests and Reactor Catastrophes . . . 56
Automobile Emissions. 56
Mobility of Elements in Soil 58
References 59

Summary

The crust of our planet is stable in respect to all elements but the radioactive and radiogenic isotopes. Their decay or generation affects the concentrations of U, Th, Pb, and K.

The element carbon one of our most important sources of energy and is converted during energy production processes to carbon-dioxide which enters in part into the

The Handbook of Environmental Chemistry
Volume 1 Part F, Ed. O. Hutzinger
© Springer-Verlag Berlin Heidelberg 1992

atmosphere and hydrosphere or is precipitated in precipitates of carbonates. In contrast to the warnings given by the Club of Rome none of the metals of our planet is used up. Only the dispersion is dramatically increased so that the amount of energy (and money) to produce the metal is much higher than before. In the future not only natural deposits will be used for mining but also anthropogenic environmental sinks such as lake- and river-sediments, sewage and municipal dumps. Proper techniques have still to be developed. Recycling of used material and production scrap will be more important. The properties of many man-made substances differ from natural substances so that the usual times of decomposition (oxidation) in soils are changed. With certain metal-organic compounds their ability to travel as complexes in water is increased. Thus new possibilities for elements to enter the food chain are opened up and additional toxic compounds may be formed.

Introduction

In 1972, the 24th International Geological Congress in Montreal, Canada organized a symposium "Earth Sciences and the Quality of Life".

In this connection the problems of exploitation of mineral deposits [1], the consumption of mineral resources [2], the deposition of wastes [3] and its implications, the importance of soil and its composition [4], and the importance of major and trace elements in soils and plants for human health [5] were addressed and discussed. Special problems were named in respect to toxic elements such as lead, mercury, cadmium, silver, tin, aluminium, gallium, germanium which were accused of being hazardous environmental contaminants, while copper, zinc, manganese, cobalt, molybdenum, selenium, chromium, iron and iodine were identified as being trace elements with a biochemical function of some kind.

At that time, the question arose of how an original natural environment is characterized, i.e. what is a normal background element spectrum.

During 1974–1979 the German science foundation initiated investigations into the "Geochemistry of environmentally important trace constituents" with reference to anthropogenic contamination [6]. It was intended, to cover the distribution of the following elements:

Sb, Be, Pb, Cd, Hg, Se, Te, Bi, As, Cr, F,
Co, Cu, Mn, Mo, Ni, Tl

As the results demonstrate, for many elements, the field of industrial and anthropogenic contamination had been ignored for years [7]. This was especially clear for lead, which, for more than 25 years, had been released in large amounts into the atmosphere from gasoline additives [8].

Cadmium was also locally enriched in the top few centimeters of soils, possibly emitted from coal-fired power plants.

Original background values from pre-industrial times can only be extrapolated from ice cores of Arctic and Antarctic snow strata. Murozumi et al. [9] calculated, that about 2% of the lead smelted in 1750 was transferred to the atmosphere and is documented in the ice. So 100000 t of lead were smelted in 1750 with an emission of 2000 t worldwide.

Natural Concentrations of Elements in the Continental Crust

The mass of the earths crust is only 0.4% of its total mass. From this amount is only 0.1% is contained in the continents. A large part of the oceanic crust consists of basalt (mostly morb but also oib) with additional amounts of marine clay sediments.

Due to differing processes of formation of the rocks and sediments in the crust, it is difficult to reach full agreement on the average composition by

Table 1. Geochemical characterization of elements

Chalcophil	Lithophil
(Cu) Ag	Li Na K Rb Cs
Zn Cd Hg	Be Mg Ca Sr Ba
Ga In Tl	B Al Sc Y La-Lu
(Ge) (Sn) Pb	Si Ti Zr Hf Th
(As) (Sb) Bi	P V Nb Ta
S Se Te	O Cr U
(Fe) Mo (Os)	H F Cl Br I
(Ru) (Rh) (Pd)	(Fe) Mn (Zn) (Ga)

Table 2. Oxidations processes under atmospheric conditions

$$As^{-3} \longrightarrow As^0, As^{+3}, (As^{5+}O_4)^{3-}$$
$$Bi^0 \longrightarrow Bi^{+2}$$
$$Br^{-1} \longrightarrow Br^{+5}O_3^-$$
$$Cl^{-1} \longrightarrow Cl^{+7}O_4^-$$
$$Co^{+2} \longrightarrow Co^{+3}$$
$$Cr^{+2} \longrightarrow (Cr^{6+}O_4)^{2-}$$
$$Cu^0 \longrightarrow Cu^{+1}, Cu^{+2}$$
$$Hg^0 \longrightarrow Hg^{+1}, Hg^{+2}$$
$$In^{+1} \longrightarrow In^{+3}$$
$$J^{-1} \longrightarrow J^{+5}O_3^-$$
$$Mo^{+4} \longrightarrow (Mo^{6+}O_4)^{2-}$$
$$S^{-2} \longrightarrow S^0, S^{+4}, (S^{6+}O_4)^{2-}$$
$$Sb^{-3} \longrightarrow Sb^0, Sb^{+3}, Sb^{+5}$$
$$Se^{-2} \longrightarrow Se^{+4}, Se^{+6}$$
$$Sn^0 \longrightarrow Sn^{+2}, Sn^{+4}$$
$$Te^{-2} \longrightarrow (Te^{6+}O_6)^{6-}$$
$$Tl^{+1} \longrightarrow Tl^{+3}$$
$$U^{+4} \longrightarrow (U^{+6}O_2)^{-2}$$

weighing the simple modes and factors. Data have been collected from various sources in Tables 2 and 3.

In Table 1 is further included the prevailing geochemical character of the elements to characterize those conditions where compounds (species) are stable. Many of the metal elements occur as sulfides in their deposits remote from the atmosphere in a reducing environment. After mining, this changes to an oxidizing environment at the earth's surface, either because the prepared metals are exposed to the atmosphere or because by-products or reject waste are disposed off.

During the development of our earth, many of the concentrations of heavy metals have been buried in the depths. Mining and consuming these metals and elements in many cases means an irreproducible spreading of those materials. Numerous examples demonstrate this continuing increase of the entropy of elements:

> boron in detergents
> cadmium in rust prevention
> lead as an antiknock additive
> platinum in catalysts

All these elements are so extensively dispersed all over our planet, that they can never (in terms of man's lifetime) be recycled.

Siderophilic elements which occur in the metal phases of meteorites are extremely rare in the earth crust. Among those elements are the Noble Metals, Cu, Hg, Pb, Sn.

In metallic form, As, Sb, Bi, Se, Te also occur. Heavy metals which today are used at the earth's surface come form the depth of the crust where they were stable mostly in the form of sulfides. Under atmospheric conditions they change to the oxidised state and are in part easily soluble and toxic. They are transported by solutions until they become fixed again. They sometimes enter the food chain.

Table 3. Average concentrations in the continental crust (in ppm)

	Wedepohl [10]	Berry and Mason [14]	Taylor [15]	Krauskopf [16]
Li	no data	30	20	20
Be	2.9	2	2.8	3
B		3	10	10
Mg	16000	20900	23300	23000
Ti	4680	4400	5700	5000
V	109	110	135	110
Cr	88	200	100	100
Mn	800	1000	950	1000
Fe	42000	50000	56300	54000
Co	19	23	25	22
Ni	45	80	75	75
Cu	35	45	55	50
Zn	69	65	70	70
As	3.4	2	1.8	1.8
Se	0.077	0.09	0.05	0.05
Zr	152	160	165	165
Mo	1.5	1	1.5	1.5
Ag	0.10	0.1	0.07	0.07
Cd	0.10	0.2	0.2	0.15
Sn	2.5	3	2	2.5
Te	0.02	0.002	no data	
Pt	0.013	0.005	no data	
Au	0.0025	0.005	0.004	0.003
Hg	0.02	0.5	0.08	0.02
Tl	0.49	1	0.45	0.8
Pb	15	16	12.5	12.5
Bi	0.08	0.2	0.17	0.15
U		2	2.7	2.7

Table 4. Magmatic and metamorphic rocks (after Wedepohl [10], Heier and Billings [17], Harder [18]) (concentrations in ppm)

	Granitic rocks	Gneisses, mica schists	Basaltic and gabbroic rocks	Granulites
Li	38		16	
Be	5.5	3.8	0.6	2.1
B	15	55	5	
Mg	6000	13000	37000	14000
Ti	3000	3870	9700	3520
V	94	60	251	73
Cr	12	76	168	88
Mn	325	600	1390	895
Fe	20000	33000	86000	38000
Co	4	13	48	15
Ni	7	26	134	33
Cu	13	23	90	27
Zn	50	65	100	65
As	1.5	4.3	1.5	(4)[a]
Se	0.04	0.08	0.09	(0.08)[a]
Zr	145	168	137	153
Mo	1.8	(1.5)[a]	1	(1.5)[a]
Ag	0.12	0.08	0.11	0.09
Cd	0.09	0.10	0.10	0.10
Sn	3.5	2.5	1.5	2.5
Te	0.01	(0.02)[a]	0.008	(0.02)[a]
Pt	0.005	(0.01)	0.03	(0.01)[a]
Au	0.0024	0.003	0.004	0.0015
Hg	0.03	0.02	0.02	(0.02)[a]
Tl	1.1	0.65	0.08	0.28
Pb	32	16	3.5	9.8
Bi	0.19	0.10	0.04	0.04

[a] Estimated concentration

Various authors have calculated the average composition of the continental crust. Depending on the model used, slight differences in the figures appear. Nevertheless, the agreement of the data is rather good especially for the trace elements.

Since the crustal rocks constitute the base for soil formation, more differentiated information is desirable: Ultrabasic, Basic, intermediate and acid magmatites are evaluated from many sources by Wedepohl [10], Seim and Tischendorf [11], Rösler and Lange [12], Paul and Huang [13] and others.

Also for clastic (graywackes, sandstone and shale) and chemical sediments (carbonates) averages have been compiled. In Table 3 only averages for the continental crust are listed from various sources.

For particular rock types, the reader may refer to the publications mentioned above.

Magmatic, Metamorphic and Sedimentary Rocks

The best recent compilation of geochemical data on the most important continental rocks and sediments is by Wedepohl (1991). He calculates the continental crust to be

- 22% granite rock
- 23% gneiss and mica schists
- 17% gabbros, amphibolites and basalts
- 37% granulites

Graywackes are mechanical disintegration products of rocks. Shales are weathering products of magmatic, metamorphic and sedimentary rocks.

Limestones have been precipitated from mostly marine solutions (environment). These rocks and sediments are the most important substrates and base for all our soils.

Table 5. Sedimentary rocks (compilation by Wedepohl 1991 [10] and Rose et al. 1979 [19]) (concentrations in ppm)

	Shales Wedepohl	Rose	Graywackes Wedepohl	Rose	Limestones Wedepohl	Rose
Li	60	66		15		5
Be	3	3	3	0,x	0.5	0,x
B		100		35		20
Mg	16000	15700	13000		26000	
Ti	4600	3800	3800		400	
V	130	130	67	20	20	20
Cr	90	90	50	35	11	11
Mn	850	850	750	x0	700	1100
Fe	4800	47000	38000	9800	15000	3800
Co	19	19	20	0.33	2	0.1
Ni	68	68	40	2	15	20
Cn	45	42	45	10	4	5
Zn	95	100	105	40	23	21
As	10	12	8	1.2	2.5	1.1
Se	0.5	0.6	0.1	0.05	0.19	0.88
Zr	160	160	450	220	19	19
Mo	1.3	2.6	0.7	0.2	0.4	0.4
Ag	0.07	0.19	0.1	0.25	0.0x	0.1
Cd	0.13	0.3	0.09	0.0x	0.16	0.035
Sn	2.5	6	(3)	0.6	0.x	0.x
Au	0.0025	0.004	0.003	0.005	0.002	0.005
Hg	0.45	0.02–0.04	0.11	0.03	0.03	0.04
Tl	0.68		0.20		0.05	
Pb	22	25	14	10	5	5
Bi	0.13	1.0	0.07	0.3	0.02	
U		3.7		1.7		2.2

The main sediments are shales, graywackes and limestones. Data from Wedepohl [10] and Rose [19] differ for the reason that different amounts of sandstones and dolomite were taken into account.

Radioactive Elements

Two types of natural radioactive elements exist one group being formed with the generation of our cosmic system and having long half-lives of a billion up to many billion years. These radioactive isotopes behave like inactive isotopes and are involved in all rock-, sediment- and soil-forming processes. During their decay radiogenic isotopes are formed which mostly have completely different chemical properties. Analysing the radioactive mother-isotope and radiogenic daughter concentrations in closed systems is often used for age determination of rock generation or dating of metamorphic events. Only a few important isotope pairs may be listed here [20].

During rock formation processes mother- and daughter-isotopes often demonstrate severe differences in chemical behavior and properties. For example Uranium is lithophile and its daughters, the lead isotopes are chalcophile. If suitable conditions prevail the radiogenic lead is incorporated into lead sulfide deposits. Since they are no longer in contact with continuously decaying uranium and thorium their lead isotope ratios (^{206}Pb, ^{207}Pb, ^{208}Pb) are fixed [21]. Lead isotope patterns have thus been determined by Bielicke and Tischendorf [22], and Doe [23].

From lead isotopes studies of rocks, minerals, and metals their original path of development can be elucidated.

Similar information can be obtained from Sr isotope [20, 24] investigations and Sm/Nd studies [25].

The other group includes 18 cosmic ray produced radio nuclides which are formed in the higher atmosphere by (n, p) reactions. Except for the three isotopes (^{10}Be, ^{14}C, ^{36}Cl) they have short half lives (Table 7).

Table 6.

Abundance	Radio nuclide	inactive daughter	Half lives/years
99.27%	^{238}U	^{206}Pb	4.510×10^9
0.7204%	^{235}U	^{207}Pb	0.7129×10^9
100%	^{232}Th	^{208}Pb	13.89×10^9
27.83%	^{87}Rb	^{87}Sr	4.88×10^{10}
0.0117%	^{40}K	^{40}Ar	1.31×10^9
62.602%	^{187}Re	^{187}Os	5×10^{10}
15.0%	^{147}Sm	^{143}Nd	1.06×10^{10}
2.6%	^{176}Lu	^{176}Hf	2.9×10^{10}

Table 7.

Radionuclide	Half-Life	Radionuclide	Half-Life
^{3}H	12.3 years	^{32}P	14.3 d
^{7}Be	53.3 d	^{35}S	87.5 d
^{10}Be	1.6×10^6 years	^{38}S	2.8 h
^{14}C	5730 years	^{34m}Cl	32 min
^{22}Na	2.6 years	^{36}Cl	3×10^5 years
^{24}Na	15 h	^{38}Cl	37.2 min
^{28}Mg	20.9 h	^{39}Cl	56 min
^{31}Si	2.6 h	^{39}Ar	269 years
^{32}Si	101 years	^{85}Kr	10.7 years

Only ^{3}H, ^{10}Be, ^{14}C and ^{36}Cl are of importance on earth in sediments, soil, plants and waters.

Soils as Links Between the Lithosphere and Biosphere

Magmatic rocks close to the earth's surface may consist of granites, granodiorites, diorites and syenites while basic magmatic rocks include various basalts, gabbros, ultramafics and ultrabasics. Sedimentary rocks can be grouped into shales, graywackes, sandstones and carbonates (and other chemical sediments).

Metamorphic rocks can be formed from both groups if the necessary temperature and pressure is applied. This may be provided by simple gravitational sedimentation, by additional tectonic stress or by submerging.

These rocks cannot provide the necessary nutrients for plants unless they are transformed into soils with clay minerals containing ion exchange qualities, organic substances (including humic compounds), and the necessary mechanical structure to allow exchange of soil–solutions and –gases. An essential step in this direction is weathering. Physical and chemical weathering in geochemical cycles was discussed in a NATO advanced study institute publication [26].

For growth, all plants need a certain amount of soil which forms from the initial rock or sediment. The development of soils depends on temperature, precipitation, topography, biological activity, and time. The soil provides the nutrients for the plants. Soil thickness varies from millimeters to 20 meters or more. Since the soils are formed from underlying rocks and sediments, they contain the same element spectrum.

Due to the additional elements from atmosphere and biosphere and due to different behavior of the newly formed components (clays, oxides, and organic substances) several horizons are generated in a soil profile. Organic carbon concentrations and elements accreted by organic substances are highest in the uppermost soil layer.

Important constituents of soils are [27]:

1. Clay minerals (Kaolinites and Halloysites, Smectites and Vermiculites, Illites and Chlorites
2. Oxides of Si, Al, Fe and Mn

    Gibbsite, Haematite, Goethite, Allophanes
    $SiO_2 \cdot x\, H_2O$, Opal, Buserite

3. Organic substances (nonhumic and humic substances, but not living matter)
4. Soil organisms (edaphon—bacteria, actinomyces, basidiomycetes, protozoa, ciliates, flagellates and others.)

These components give rise to physical and chemical conditions within the soil profile arranged in horizons of different composition. Due to the various properties of the above mentioned components inorganic elements are distributed accordingly.

Since both the climatic conditions and the substrate differs from place to place, various soil profiles have developed and they have been defined and mapped by FAO -UNESCO [28] and by Tavernier and Louis [29].

Widely used modern systematics in Europe have been established by Schachtschabel et al. [27].

Almost all the nutrients of the world originate from the thin veneer of soil on our planet and this thin veneer provides all the elements and components plants need.

Although it is difficult to give representative concentrations for elements in soils, various examples are listed from the literature (Table 8).

Table 8. Trace elements in uncontaminated soils (ppm)

	Bowen [30]	Andrews-Jones [31]	Seto & Deangelis [32]	Brüne [33]	Müller et al. [34]	Essington & Mattigod [35]	Schöttle et al. [36]
Ag	0.1	1					
As	6	5	6.5	9			
Br		10					
Cd	0.06	0.5	0.7	0.1	≤ 0.31		
Co	8	10	4.5		≤ 7.2	17.8	
Cr	10	200	14	39	40.4	43.2	16.6–52.4
Cu	20	20	25	18	15.7	46.6	5.0–38.9
F							
Hg	0.03	0.01	0.08	0.09	0.134		
Mn	850	850			640	1009	
Mo	2	2.5	0.4				
Ni	40	40	16.0	38	19.6	38.8	6.5–88
Pb	10	10	14.0	25	40.6	24.7	11.4–50.3
Se		0.5	0.4				
U							
V	100	100				88	
Zn	50	50	54	66	80.2	93	29.7–119.8

Use of the Pedosphere in Geochemical Exploration for Deposits

Geochemical exploration, in many cases, uses soil sampling for finding ore occurrences. Since visible outcrops of mineralizations are rare after many hundred years of prospecting and mining the "natural contamination" of soils during weathering and soil formation is used as an indicator of mineralization. For any geochemical prospecting work the local background has to be determined for the element of interest. The normal abundance of an element in unmineralized earth is commonly regarded as background. The background values of different materials differ considerably from each other. Rose et al. [19] give general background levels for 30 principle minor elements in normal rocks.

A geochemical anomaly is confined by a set of data which indicate concentrations above the upper limit of normal background fluctuation. The boundary between background values and anomalies is called the threshold. It can be of regional or local importance. Sometimes it is easier to look for associated elements – so called "pathfinders" – and not for the major components themselves.

Different ore types are characterized by certain element associations which can be used to get indications of the mode and conditions of formation. Some associations are listed below (Table 9). Normally those elements used for prospecting purposes are those which are easiest to detect and determine and the integration of data is clear. Either soils themselves or waters or gases can be used if sensitive measuring techniques are applied.

Newer techniques with a minimum of sample preparation for soil surveys are described by Kramar and Puchelt [37, 38] (energy dispersive XRF) and van den Boom and Pöppelbaum [39]. Informative examples on geochemical exploration projects are given by Fletcher et al. [40]

Systematics of the Pedosphere

Normally, the pedosphere is, under European climatic conditions, divided into various soil-horizons. When they consist of organic matter ($\geq$ 30% organic substances) they are designated by H (histic) if they consist of peat and undecomposed litter by O (organic) or L (litter) which rests on top of the mineral soil.

A   is the uppermost mineral horizon in the upper soil (top-soil) and contains humic substances or is depleted in minerals

E   eluvial horizon depleted in humic substances, clay, iron and manganese compared to A

B   is the mineral horizon in the subsoil with altered mineral composition by introduction of minerals and substances from the topsoil or by weathering in situ. Solid residues of the parent rock are less than 75% and normally the colon is changed compared to the parent.

G   is the characterization of a mineral horizon at the ground-water level
lc   is loose parent rock with fine soil
mc   is hard rock.

More criteria may characterize the rock by small letters placed in front:

a   alluvial ($\geq$, 0, 35% organic substance or sulfide content within a depth of
     125 cm
o   aeolic substrate, wind deposited
q   siliceous rock ($\leq$ 2% $CaCO_3$)
e   marly rock (2–75% $CaCO_3$)
k   limey rock ($\geq$ 75% $CaCO_3$)
t   clay rock ($\geq$ 45% clay)
f   fossil horizon
r   relict horizon

Small letters after the soil horizon designation describe the characteristics of soil
developing processes.

Definitions are given in Schachtschabel et al. [27] and FAO-UNESCO [28].
Soil in Germany are categorized by the Arbeitsgemeinschaft Bodensystematik
[41]. The World Soil Map from FAO and UNESCO scale 1:5 Mill. uses a new
international nomenclature with 25 soil units corresponding to the US system.

Geochemical Distribution of Elements in Soil Horizons and Catena

The average element content of soils is not equally distributed over the horizons
of our stratified soil. Elements are either contained

- in parent material minerals
- in authigenic soil minerals of secondary origin
- adsorbed to compounds with suitable exchange sites (clays, oxides, hydrox-
  ides, humic sustances)
- coprecipitated with newly formed minerals
- contained in living or dead organic matter
- in soil solutions

Transfer into plants can only occur via dissolution. The solvent is always
water, mostly using the aid of complexing (organic) compounds and protons
(pH). Not the total concentration of an element in soil is relevant for plant
growth but rather the fraction which is plant-available. This is due to the mode
of bonding and the specification of the element in addition to the prevailing pH
and the presence of (natural) complexing compounds. Chemical aspects of soils
were discussed by Paul and Huang [13] in this series. Role and importance of
adsorption, cation exchange complexation, stability, constants and specification

Table 9. Path finder elements in soil for various deposits [19, 42, 43, 44, 45]

Path finder elements (associated elements)	major element	type of deposit
As Sb Hg S Te	Au Ag	epithermal precious metal
Ag As Sb	Au	hydrothermal vein quartz
Ag As S	Cu	Copper in basalt
(Cu) Co (Ni) Zn	Cu Ni	
As Mo Zn	Cu	porphyry copper
Ag Cd Co Hg Ni Pb Zn	Cu	copper shale
Ag As Ba F Hg	Co Ni (Bi)	hydrothermal vein
(Ba) Cu (F) Hg Pb	Ba F	hydrothermal or synsedimentary
Ba Pb Zn	F	fluorite Mississippi valley type
F Li B	Sn W	greisentyp
Hg	sulfides (Pb Cu Zn)	sulfide deposits
Tl	Fe (S) Zn	volcanic exhalative
(Be) Cu (W) (Mo)	W Mo	scarn type mineralization
V	U	colorado type
Cu Mo Pb Se V	U	sandstone type
Ag Co Ni	Cu	copper sandstone

Table 10. Nutrients and pollutants in soil

the first group contains	the second group
Ca	SO$_2$
Mg	F
K	Cd
Na	Pb
P	Hg
N	Ni
Mn	Cr
Fe	As
Cu*	
Zn*	
B*	
Mo*	
Cl	
Co*	

and the kinetics between different situations are described by Schmitt and Sticher [46]. Underwood [47] links soil via plants and animals to the food chain.

Schachtschabel et al. [27] group elements into nutrients and pollutants (toxic agents).

From the first group those elements marked with an asterisk can be regarded as necessary nutrients only up to certain concentrations. Above this limit they turn into pollutants. Due to their different chemical characters, the elements are enriched in certain soil horizons. The fact that Be, Co, Ni, (Zn), Ge, As, Sn, Tl, Ag, Au are enriched in the organic horizon of soils (Fig. 1) was found out by Goldschmidt as long ago as 1937. The fact that he also grouped Pb and Cd under the biophile elements according to his findings was probably caused by atmospheric transport from smelters.

A latosol profile from Zambia (Fig. 2) demonstrates a clear decrease of Cu, Cr, V and (Fe) in the uppermost (probably organic rich) part of the sequence (Higher concentrations are found in the B horizon, were clay minerals and manganese— and iron oxides are accumulated [19].

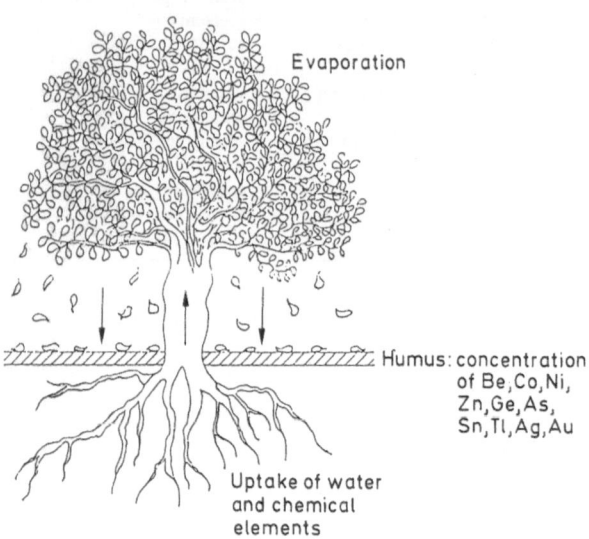

Fig. 1. Uptake of chemical elements from the soil by plants and concentration of the elements in the uppermost soil horizon [48]

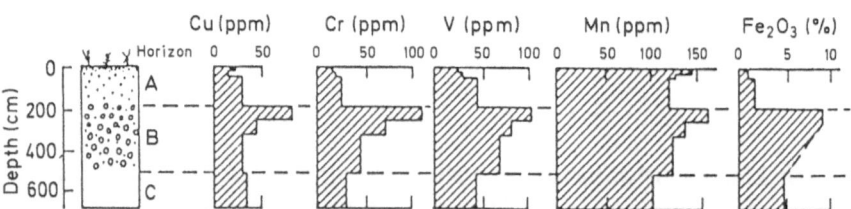

Fig. 2. Variation in metal content with soil horizon, latosol profile, Zambia. Alkalis increase with depth; Co and Ni show little change. Data on − 80-mesh fraction. (Sampling by J. S. Tooms; analyses by J. D. Kerbyson, Geochemical Prospecting Research Centre, Imperial College, London.)

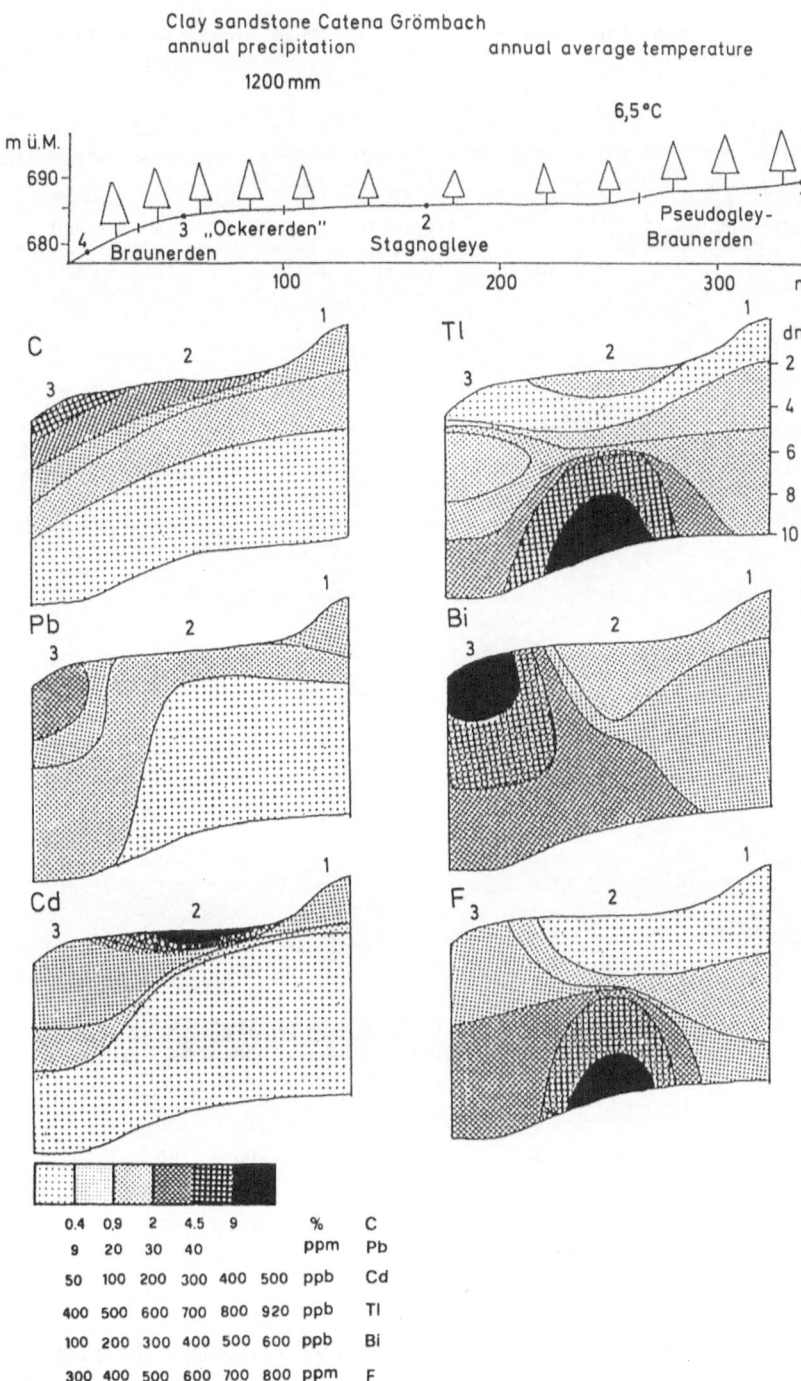

Fig. 3. Clay sandstone Catena Grömbach

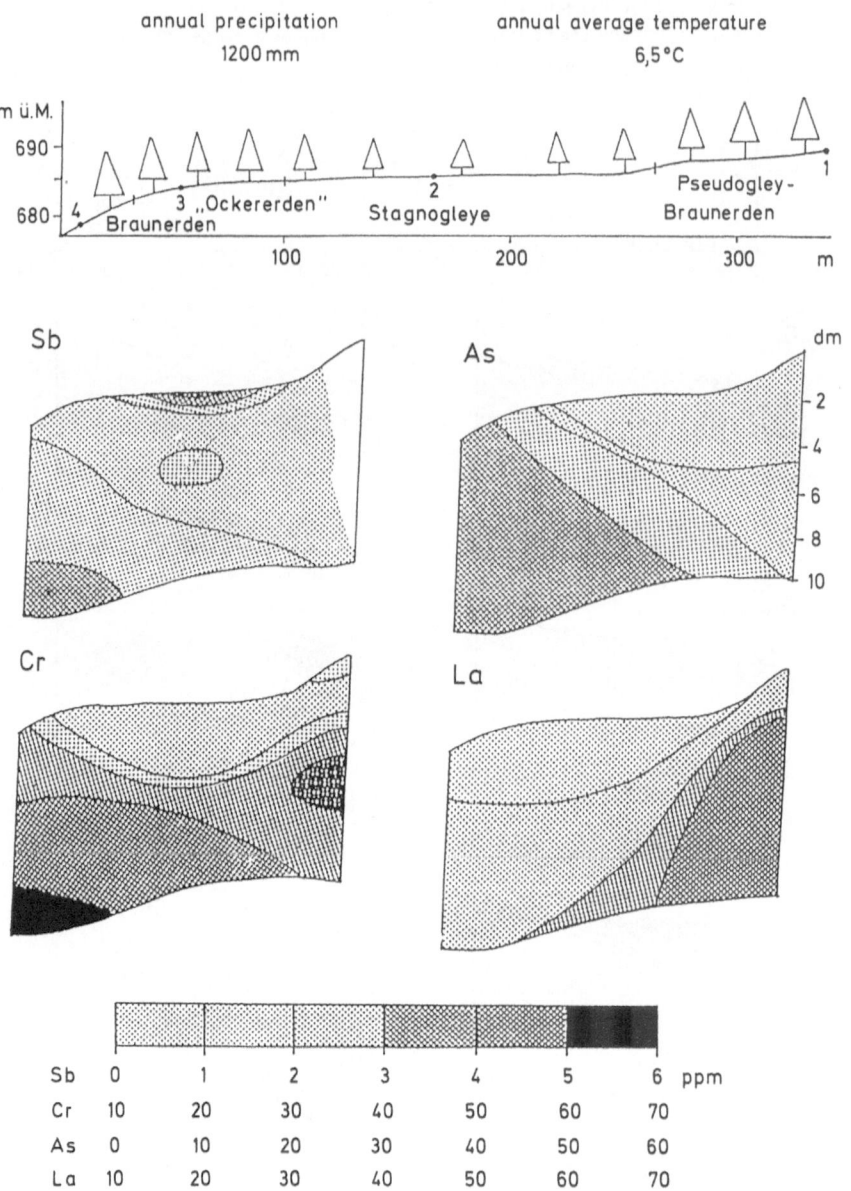

Fig. 4. Clay sandstone Catena Grömbach

The situation in soil development on "Bunter" (Triassic) in the Black Forest, Germany under different morphological conditions is shown in Figs. 3 and 4. In these figures lead and cadmium are enriched in the organic layer indicating some anthropogenic (atmospheric) pollution [49].

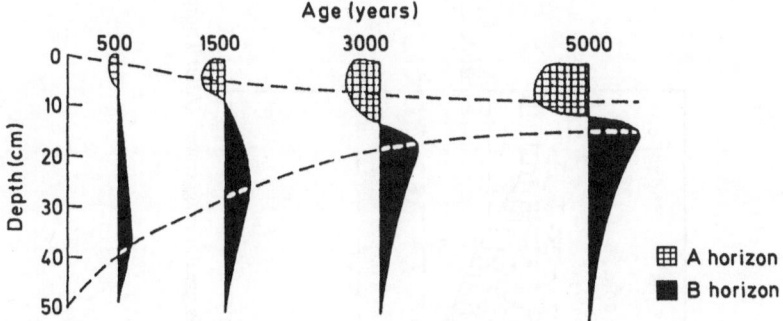

Fig. 5. Formation of A and B horizons of a podzol profile as a function of age. (After Jenny)

Formation of soils with well-developed horizons takes several hundreds to thousands of years as shown by Jenny for a podzolic profile [50] (Fig. 5).

Influence of precipitation and morphology is visible from a series of soil profiles from Wyoming [51] (Fig. 6)

Upper Limits for Concentrations of Elements in Soils for Agricultural Use

Continuous contamination of soils by toxic elements dispersed by industry, traffic and other anthropogenic effects requires measures to protect people from hazards to their health. In order to avoid introducing metalloids and heavy metals into food stuff via uptake from soil, various ordinances have been issued in American and European countries and these set upper limits for these substances in soil and soil-improving agents (fertilizer, compost, sewage sludge). The approach of the guide lines is different in different countries: While in the Federal Republic of Germany only "tolerable concentrations" of soils are listed [52] regardless of soil composition, clay and organic substance is taken in to account in the Netherlands [53]. If a certain amount of a toxic substance is soluble – which means that it is accessible for plants – the threshold for agricultural use is exceeded according to Swiss [54] and Hamburg Ordinances [55].

California uses also both thresholds in soil and for soluble concentrations [56]. British guidelines for contaminated soils [57] group them into five categories:

I   uncontaminated soils
II   slightly contaminated soils
III contaminated soils
IV   heavily contaminated soils
V   very heavily contaminated soils.

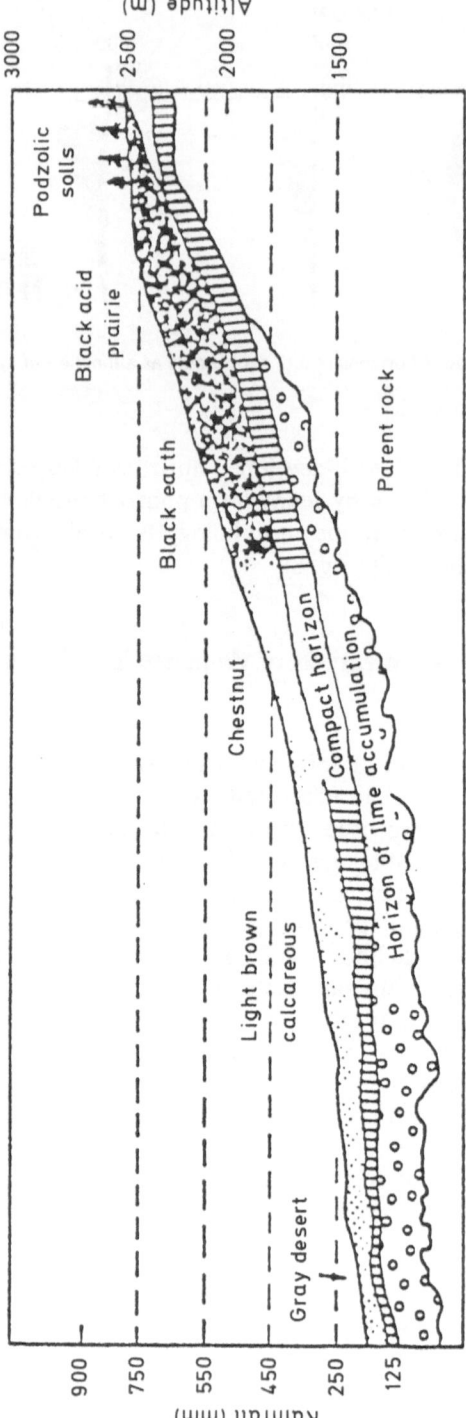

Fig. 6. Gradation of soil types from desert to humid mountain top, west slope of **Big Horns, Wyoming**. Scale of soil profiles is greatly exaggerated. (After Thorp)

Standards, guidelines and legislative regulations for metals and their compounds for agricultural soils, ground water, compost and sewage sludge have been listed and compared by Ewers [58].

Zielhuis [59] detailed the risk assessment of human exposure to metals and their compound via foodstuffs including drinking water.

From Table 11, the reader can conclude that

- the tolerable concentrations of metals in soils be supplemented by
- mineralogical composition and the concentration of (natural) organic matter,
- the soluble concentration of a compound and the pH of the soil is highly desirable information.

Landesanstalt für Umweltschutz (The State Institute for Environmental Protection) Baden Württemberg FRG [60] edited, in 1989, a comparison of guidelines and thresholds for environmental media – air, water, and soil.

Herms [61] related tolerable concentrations of heavy metals in soils to their pH. This takes into account the fact, that the solubility of most heavy metals increases with decreasing pH.

Another proposal for guidelines was made taking into account the various uses of an area [62]. Although not yet law, two models for the use of contaminated soils in urban areas and for the production of crops are given.

A critique of the geoscientifically unsatisfactory guidelines for soils containing heavy metals was published by Kuntze et al. [63] and they define the problems and point out some possible solutions.

Anthropogenic Contamination of Soils with Inorganic Substances and Their Distribution in Soil Horizons

Elements are distributed in all soils due to the weathering of their parent rocks. Their concentration is the natural geogenic background. In addition human activity caused deposition on or infiltration of toxic compounds into the soils. This is due to ignorance, carelessness or poor technique. As the reactions of nature to this pollution are investigated more and more causes are being identified.

Emissions from Industrial Sources

One important source of industrial emissions and probably the oldest contaminator is the mining industry and its connected metallurgy. In the United States a list of 952 sites are reported as cases of damage caused by metals and metaloids [63].

Table 11. Various guidelines on concentrations of metals in soils

	BRD [52] tolerable concentr. ppm	Netherlands [53] A normal	Netherlands [53] B further investigation	Netherlands [53] C sanitation necessary	Switzerland [54]	Hamburg [55] further investigation	Hamburg [55] sanitation necessary	California [56] dangerous concentr.	California [56] very dang. concentr.	EG [57] slightly contam.	EG [57] contam. soil	EG [57] heavily contam.	EG [57] very heavily contam.
Aluminium	–												
Antimony	5												
Arsenic	20	29	30	50		50		500	50000	30–50	50–100	100–500	≥ 500
Barium	–	200	400	2000		400	2000						
Beryllium	10												
Cadmium	3	0.8	5	20	0.8	8		500		1–3	3–10	10–50	≥ 50
Chromium	100	100	250	800	75	300		CrIII 8000 CrVI 2000		100–200	200–500	500–2500	≥ 2500
Cobalt	50	20	50	300	25	50	300						
Copper	100	36	100	500	50	300							
Gallium	50												
Lead	100	85	150	600	50	300		1000		500–1000	1000–2000	2000–10000	≥ 10000
Mercury	2	0.3	2	10	0.8		5			1–3	3–10	10–50	≥ 50
Molybdenum	5	10	40	200	5	40							
Nickel	50				50	300	200	100	10000	1–3	3–10	10–50	≥ 50
Selenium	10												
Thallium	1				1								
Tin	50	20	50	300		50	300						
Titanium	5000												
Uranium	5												
Vanadium	50												
Zinc	300	140	500	3000	200	1000		5000					
Zirconium	300												

Mining Industries and the Connected Metallurgical Industry

Lead mining and smelting was known and practised 5000 years ago in SW Asia. Later on silver which is often associated with lead mineralization was the main aim of mining and metallurgy had to remove the unwanted lead by oxidation. Murozumi et al. [9] showed, by analyzing ice cores from the Greenland and Antarctic snow strata, that from the industrialization in A.D. 1750 a continuous increase of lead in the atmosphere could be observed. They found the unpolluted background level of arctic air (snow) for 800 B.C. which indicates that lead emitting metallurgy existed prior to the eighteenth century. Recent techniques of metallurgy and refining are described by Wilmoth et al. [64].

Mining produces valuable ores and unwanted byproducts and gangue. The worthless components have been deposited mostly in dumps close to the mining site. In this century ores have been mostly ground to fine particles so that separation of the valuable constituents could be achieved by flotation.

In the case of silver mining, the worthless byproducts often contain arsenic, antimony, zinc, copper, cobalt, nickel and sometimes thallium and cadmium.

Only a few examples will be given:

At Mechernich there is a lead ore deposit which was mined by the Romans in the second century A.D. Mining activities continued there until 1957. Altogether 1.2 million tons of lead were produced over the years. Most of the ore was smelted close to the mine or nearby. In order to reduce the possibility of lead poisoning of the smelter workers the height of the smelter chimney was increased to 134.6 m. This also reduced the amount of flue dust.

The 1.2 million tons of lead was produced from 60 million m^3 of ore. Schalich et al. [65] described the history of this mining enterprise and the lasting consequences for the area. Heavy pollution of the soil with Pb concentrations of up to 5000 ppm was the result. Large mining dumps are in part not yet stabilized. Anthrapogenically affected soils contain ca 1000 ppm Zn, while Cd is in most cases below 3 ppm. One additional problem in this area is the transport of flotations products by wind and water.

Another example is the deposit at Wiesloch, south of Heidelberg, which was mined for 2000 years until 1954 [66–67]. It contained silver, lead and zinc together with thallium, cadmium, arsenic and antimony as sulfides, carbonates or oxides.

Since the last named elements had no commercial interest they went to the local or central mine dumps. Local mine dumps often accumulated close to the shafts or adit. When the particular shaft was abandoned and the area was reclaimed for agriculture, the dumps with all their contents were leveled.

Arsenic, cadmium, thallium, and antimony were thus incorporated into the soil where they locally inhibited the growth of the crop. Investigation of such "infertile" spots clearly showed heavy contamination by toxic elements (As up to 7200 ppm, Cd upto 115 ppm, Tl upto 428 ppm, Sb upto 128 ppm, Zn up to 47 000 ppm, Pb upto 10 800 ppm [68–69]

One of the spots where agricultural growth of grain, corn, and rape was markedly reduced has been carefully investigated in soil profiles down to 150 cm

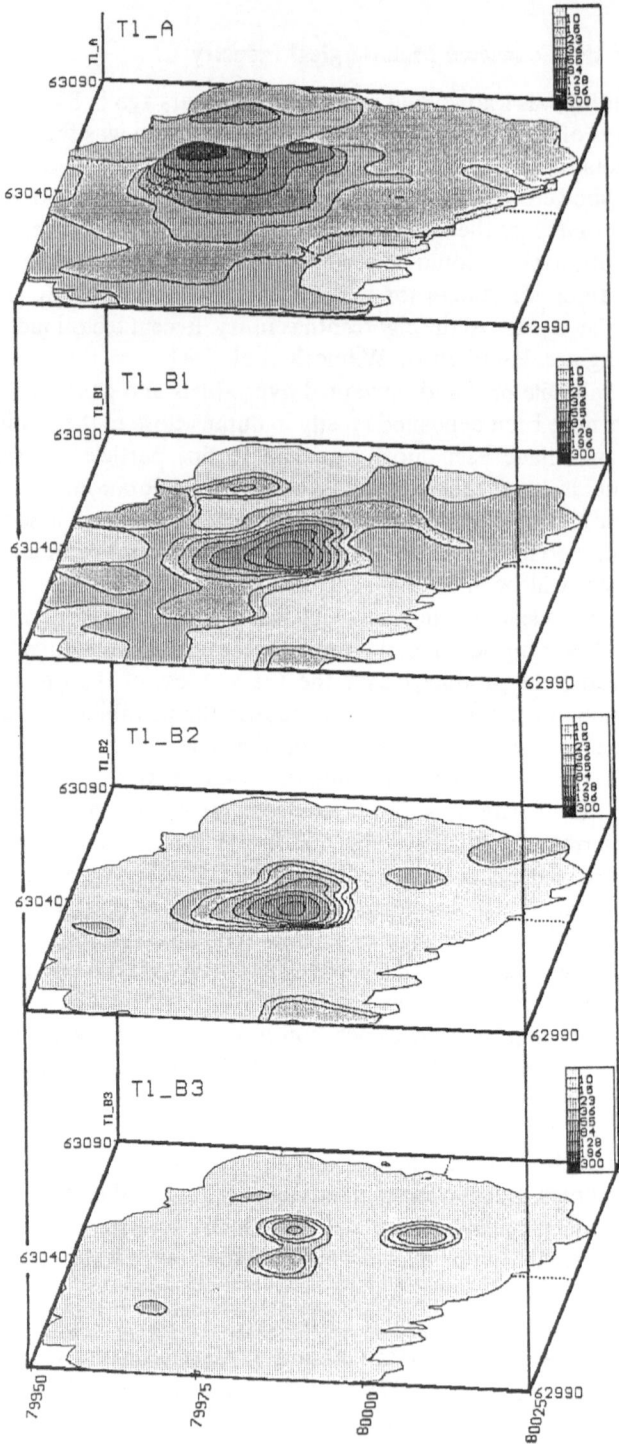

Fig. 7. Levels of Thallium in a field near Wiesloch, FRG., contaminated by mining activities.

into the loess. From the analytical data it is clearly visible that the contamination originated from mining dumps on the surface, from which it penetrated into the soil and sediment. It cannot be decided which of the eolic co-occurring elements As, Cd, Pb, Se, Tl, Zn are responsible for the depression of growth or whether all of them operate synergetically.

In Fig. 7 one example is given for thallium contamination in soil horizons from a paper by Puchelt et al. [68]. Such heavily polluted soils also affect the plants still growing next to maximum contamination. They incorporate the toxic elements differently into the parts of the plants and often discriminate them from the seed proper [140].

Thallium concentrations are given in ppm for the following soil profile depths

0–35 cm	:	Tl	A
35–80 cm	:	Tl	B 1
80–100 cm	:	Tl	B 2
100–150 cm	:	Tl	B 3

On the x and y axes the geographic position is given in Gauss–Krüger coordinates.

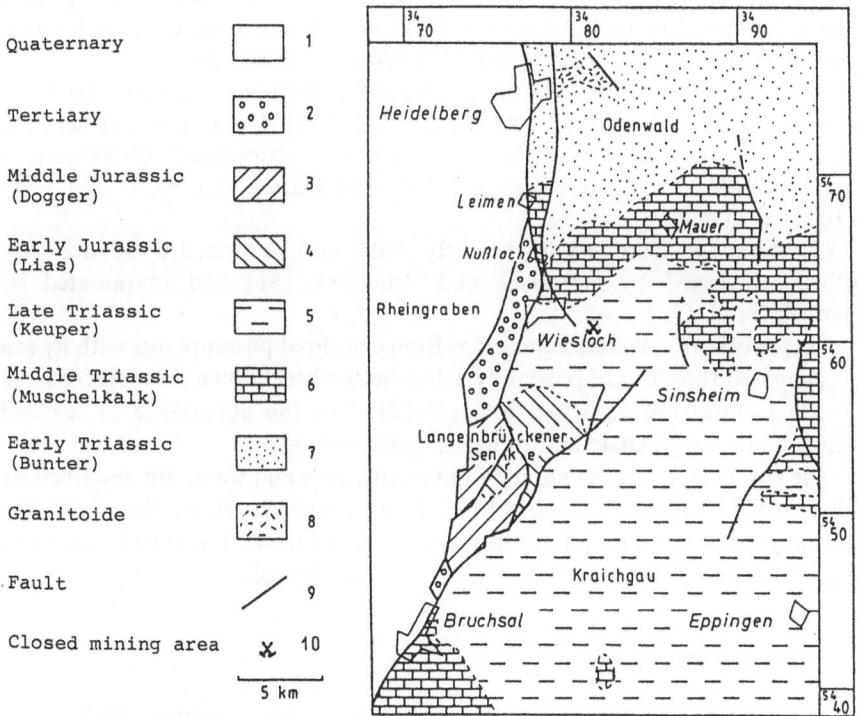

Fig. 8. Geological map of Wiesloch area with locations of former mining activities

The same situation has been found in several old mining locations in the Black Forest where mostly arsenic, antimony, sometimes copper, lead and uranium contaminates the soil [70–72]. Since the last century, contamination by arsenic from mine dumps and smelters has been known to exist in the area north of the Harz mountains [73].

A cause of heavy contamination of soils are the milling sands of medieval times. They contain considerable amounts of heavy metals and are transported each year with melt water and floods [74]. This book also reports on prehistoric and ancient mine dumps, contaminating the surrounding area.

Smelters and refineries emit large quantities of metals daily. A zinc smelter in Palmerton, USA contaminates its neighborhood with 6000–9000 kg Zn, 70–90 kg Cd, 90 kg Cu and 90 kg Pb each day [75]. A zinc smelter in Datteln, Germany emits 0.01% of its annual production through the stack; that is 11 260 kg Zn per year [76].

Further examples from other parts of Europe and from North America are listed by Ernst and Joossee van Damme [74] and Freedman and Hutchinson [77].

Coal-Fired Power Plants

The world coal production was 3600×10^6 tons in 1978. Production for the year 2000 is estimated to be 9300×10^6 tons. The cumulative production tonnage from 1979 to 2000 is estimated to be $151\,000 \times 10^6$ tons [78].

In 1980, the US National Committee for Geochemistry gave average trace element concentrations for the US and the world [79]. The amounts of elements mobilized by the year 2000 will be 14 000 tons of mercury and 9 700 000 tons of fluorine. Table 13 gives terrifying quantities for Cd, Pb, Zn, Se, U and 1.695 $\times 10^9$ tons of sulfur.

Conditions in Germany in the early 1980s were described by Kautz [80], critically reviewed by Heinrichs and Brumsack [81] and reevaluated by Heinrichs et al. [82].

These authors calculated emissions from coal-fired power plants with fly ash recycling and the effect of partial fly ash collector breakdown. The problems of fly ash disposal are discussed and compared to the abundance of selected elements in average shale and in urban sewage sludge.

The effects of fly ash on animals and plants, soils and water are described by Ernst and Joosse van Damme [74]. It is absolutely clear, that both the enormous amounts of mobilized trace elements and the fly ash must be disposed of safely so that they are not in contact with the biosphere.

Cement Factories

Cement plants use high temperatures in their furnaces ($\geq 1500\,°C$). For certain types of cement in addition to limestone and marl ferric oxides have to be used.

Table 12. Average concentrations of elements in coal

Element	U.S. average Concentration (%)	Worldwide average
Sulfur	2.0	2.0
Phosphorous	—	0.05
Silicon	2.6	2.8
Aluminium	1.4	1.0
Calcium	0.54	1.0
Magnesium	0.12	0.02
Sodium	0.06	0.02
Potassium	0.18	0.01
Iron	1.6	1.0
Manganese	0.01	0.005
Titanium	0.08	0.05
	Concentration (ppm)	
Antimony	1.1	3.0
Arsenic	15	5.0
Barium	150	500
Beryllium	2.0	3
Bismuth	0.7	5.5
Boron	50	75
Bromine	2.6	—
Cadmium	1.3	—
Cerium	7.7	11.5
Cesium	0.4	—
Chlorine	207	1000
Chromium	15	10
Cobalt	7	5
Copper	1	15
Dysprosium	2.2	—
Erbium	0.34	0.6
Europium	0.45	0.7
Fluorine	74	—
Gadolinium	0.17	1.6
Gallium	7	7
Germanium	0.71	5
Hafnium	0.60	—
Holmium	0.11	0.3
Iodine	1.10	—
Lanthanum	6.1	10
Lead	16	25
Lithium	20	65
Lutetium	0.08	0.07
Mercury	0.18	0.012
Molybdenum	3	5
Neodymium	37	4.7
Nickel	15	1
Niobium	4.5	—
Praseodynium	2.7	2.2
Rubidium	2.90	100
Samarium	0.42	1.6
Scandium	3	5
Selenium	4.1	3
Silver	0.20	0.50
Strontium	100	500
Tellurium	0.1	
Terbium	0.1	0.3
Thallium	0.1	—

Table 12. Continued

Element	U.S. average Concentration (%)	Worldwide average
Thorium	1.9	—
Thulium	0.07	—
Tin	1.6	—
Tungsten	2.5	—
Uranium	1.6	1.0
Vanadium	20	25
Ytterbium	1	0.5
Yttrium	10	10
Zinc	39	50
Zirconium	30	—

Table 13. Amounts of elements mobilized as a result of coal production in short tons[1]

Element	U.S. 1978	World 1978	Cumulative 1979–2000
As	3 300	19 000	770 000
Be	910	5 100	210 000
Cd	91	510	21 000
Co	2 270	13 000	530 000
Cr	6 110	34 000	1 400 000
Cu	7 800	43 000	1 800 000
F	42 000	230 000	9 700 000
Hg	59	328	14 000
Li	5 400	30 000	1 300 000
Mn	16 000	91 000	3 800 000
Mo	1 200	6 500	270 000
Ni	4 700	26 000	1 100 000
Pb	4 600	25 000	1 000 000
Sb	380	2 100	88 000
Se	1 100	6 200	250 000
U	850	4 700	200 000
V	9 800	54 000	2 300 000
Zn	9 100	51 000	2 100 000
S	7.34×10^6	40.9×10^6	$1 695 \times 10^4$

[1]One short ton is 0.907 metric tons

From the 1960s and 1970s, in some plants, Fe_2O_3 was introduced which originated from pyrite from Meggen/Sauerland and had been used previously for sulphur dioxide production by burning at intermediate temperatures. The pyrites from this deposit contained about 300 ppm of thallium. It was not volatilized during this step, but remained in the residual oxide. Thallium reaches its boiling point at 1460 °C and is thus volatilized. These temperatures are surpassed in cement furnaces and the element is transformed into a gaseous state. From this it can be regained only after its condensation by electrofilters or waste-gas washing. Other elements also volatilize at these temperatures: cadmium, lead, bismuth, zinc under certain conditions, selenium, tellurium. A high temperature is also an elegant analytical method of separating small concentrations of these elements from uncontaminated or contaminated soils [83]. In Wiesloch, Puchelt and Walk [84, 85] were able to demonstrate, that the cement industry was not the reason for thallium contamination but previous mining activity. The connection between the cement industry and thallium pollution of soils was clearly demonstrated by Schoer [86] and Lehn and Schoer [87].

The case from a cement plant in Lengerich/Westphalia was the first observation of this kind and led to investigations of soils around all cement plants in Germany.

Bambauer and Schäfer [88] published analyses of ferric oxide, which was used in the cement furnace at Lengerich. Its thallium concentration was 290 ppm.

Schoer and Nagel [89] correlated the thallium content in soils to the distance from the cement plant stack.

A further effect influencing the area was that, in former times, there was emission of considerable amounts of dust from cement plants. This causes an increase of pH of the soil, which made growth difficult for some plants [74].

Fluorine-Emitting Industries

Fluorine occurs in soils at concentrations from 90–980 ppm [90]. In power plants, coals are used, which contain different amounts of clay minerals which finally form ash or slag. Clay minerals and apatites contain the highest fluorine concentrations in this sytem. Fluorine reaches a concentration of 1600 ppm in hard coal. Since combustion occurs at 1600 °C the element escapes and possibly reacts with water to form hydrofluoric acid [91]. The same process occurs in the production of bricks and tiles [92]. Fluorine emission is also observed in the vicinity of hydrofluoric acid and fluorocarbon producing plants [93].

Various industries emit fluorides which accumulate around the point of emission in the soil [94]: glassworks and aluminium works [95, 96]; phosphate fertilizer plants [97–99].

Mercury-Emitting Industries

McNeal and Rose [100] assumed that approximately 50% of the annual mercury production of the world is lost to the environment during its use, i.e. 5300 t for 1971. Since that time the production and use of Hg has been reduced so that in 1982 only 6600 t were produced [101].

The main applications of mercury are:

electrical (mainly batteries)	56%
agriculture (fungicides, insecticides)	14%
electrolysis (alkali-chloride production)	13%
physical and medical instruments	10%
other applications (dental, amalgamation)	7%

In some of these applications the substances are not meant to be recycled (fungicides, pesticides, pharmaceutical products). In addition, processing of sulfide ores in smelters discharges between 1800 and 30 000 t per year through the stack [102].

Fossil fuel combustion emits, despite electrical fly dust precipitators, considerable amounts of mercury. Heinrichs et al. [82] calculated that, for the West German consumption of 82×10^6 t for coal in 1981, an emission of 22 000 kg of Hg. McNeal and Rose [100] found that mercury from the air is precipitated by rain to the soil, where it is fixed in the uppermost layers.

Another source of mercury emission is cement and brick production for which Weiss et al. [103] assume that 100 000 kg Hg is emitted each year.

Gold amalgamation poses a serious ecological problem in developing countries with small scale mining [104] since mercury is handled and distilled in a risky way by laymen.

Contamination by Fertilizers, Mineral Fertilizers, Sewage Sludge, and Compost

Phosphate fertilizers introduce large amounts of iron (5700–11 800 ppm), chromium (39–92 ppm), nickel (23–39 ppm), zinc (42–108 ppm) and cadmium (2.1–9.3 ppm) to agricultural soils [105].

Urea, ammonium nitrate, potash, potassium sulfate and agricultural dolomite supply considerably smaller amounts of these elements. Webber [106] traces high concentrations of cadmium in crops back to Pacific island phosphate fertilizers. Williams and David [107] observed a considerable increase in the cadmium content of topsoil after heavy use of superphosphate. The cadmium appeared to be readily plant-available.

Sewage sludge is a product of waste-water treatment plants and occurs in direct relation to the amount of sewage. It contains considerable amounts of organic matter and additional phosphorus if phosphate removed is included [108].

In Germany in 1989, 42 million m^3 sewage sludge with 5% dry matter was produced. In 1986, 29% was used as fertilizer in agriculture, 59% was deposited in dumping grounds, 9% burnt, and 3% composted [109].

When applying sewage sludge for agricultural purposes in Germany, the Sewage Sludge Ordinance (1982) has to be observed and this defines upper limits for concentrations of heavy metals:

Pb	1200 mg/kg
Cd	20
Cr	1200
Cu	1200
Ni	200
Hg	25
Zn	3000

A total of 5 t dry weight of sewage sludge can be applied to a 1 ha ($10\,000\,m^2$) area within 3 years.

Due to unresolved problems the Ministry for the Environment (FRG), in September 1988, recommended that sewage sludge should not be applied to grassland and for growing field crops. Förstner and Stiefel [110] found nickel and chromium introduced into sludge by galvanizing firms. Another municipal refuse is town-waste compost. Only 3% of the town waste of Switzerland was composted in 1980 [111].

Häni [114] surveys the literature on the matter. Hermite and Ott [112] edited the Proceedings of an International Sewage Sludge Symposium (1984).

Contamination by Pest Control Agents
(Herbicides, Fungicides, Insecticides)

By developing agriculture to monocultures, plant protection agents have had to be used in order to maintain high yields. For a comparatively long time arsenic has been used to fight insects. Copper arsenite has been used since the middle of the last century against the colorado beetle. Despite the development of organic pesticides, arsenic compounds are still applied in considerable quantities – As III and As V oxides are still used in cotton plants. During reforestation of an arsenic treated potato area, growth of fir trees (*Tsuga canadensis*) was heavily retarded [113]. Although use of arsenic-containing pesticides is restricted or totally banned by many countries due to persistence of the compounds, the problem is still of immediate interest.

Arsenic application caused severe contamination to agricultural land, with As residues up to 120 ppm in Canada [114] and 830 ppm in the U.S. [115]. Since arsenic behaves like phosphate in the soil it forms salts of low solubility. In order to control insects, large amounts of As ($1000-10\,000$ kg) are used per square km.

Copper has fungicidal properties. It was used as early as the beginning of last century against wheat black rust. In a mixture with calcium hydroxide (bordeaux mixture) it was often used against mildew in viniculture. Since that time, the rigorous application of copper containing biocides has led to high contamination levels in vineyards [116], hop cultures [117], potato acres and fruit plantations [118]. In a field used continuously (43 years) for hop production, the soil contained 525 ppm Cu. Earth worms react sensitively to copper stress [119].

Heavy metal contamination has been reduced since 1930 when metal dithiocarbamates were introduced as biocides. At the same time additional metal compounds (Zn, Mn, Fe and Sn) have been used and supported by metal sulfates and chlorides [120]. Another, not unproblematic, substance in agriculture is methyl bromide (CH_3Br) which is used against fungi and nematodes [121]. The substance decomposes in the soil leaving bromine ions which cause phytotoxic effects even at 5 ppm. One of the earliest agrochemicals was mercury. It is extensively used for treating seeds. By this method up to 1 mg Hg per m^2 can be introduced into the soil [122]. While in former times mercury chlorides ($HgCl$, $HgCl_2$) were used, nowadays phenylmercury acetate is mostly applied. The persistence of mercury in the soil caused very high pollution in the Netherlands (up to 10 ppm) [123]. These heavy contaminations have serious effects on birds and small mammals.

The topic of agrochemicals was widely discussed by Ernst and Joossee-Van Damme [74] and was also mentioned by Webber [106]. Freedman and Hutchinson [77] list 11 metal-containing pesticides recommended in Ontario for various fruit crops.

Contamination by Atomic Bomb Tests and Reactor Catastrophes

Additionally to the usual U decay processes, there occurs the so called "fission" which happens spontaneously with a half life of 8×10^{15} years. Its products are approximately of roughly equal mass.

$$^{235}U + n \rightarrow {}^{144}Ba + {}^{90}Kr + 2n + energy$$

These are the processes anthropogenically initiated in atomic bombs and in nuclear reactors. A world wide list of nuclear reactors was published by Koelzer [124], Eisenbud [125] lists 423 atmospheric nuclear weapon tests by the US, United Kingdom, France, Soviet Union and China. He also reports 12 accidents that resulted in contamination of the environment.

Estimated quantities of radioactive materials released from the Chernobyl reactor were published in the USSR (1986).

From Table 14 it must be concluded that ^{90}Sr, ^{125}Sb, ^{134}Cs, ^{137}Cs will be the longest living irradiation sources from the Chernobyl accident.

For Germany consequences of the reactor accident in Chernobyl including soil contamination have been measured and documented by the Strahlenschutzkommission [126].

Automobile Emissions

Automobile fuel is one of the most serious contaminants of air and soil because of its content of lead alkyls to improve the octane rating. This use of lead started on a large scale in approximately 1950 [9] and made lead contamination worse

Table 14. Estimated quantities of radioactive materials released from the Chernobyl reactor (USSR, 1986)

Nuclide	Releases (megacuries)		% of inventory released as of May 6, 1986
	April 26, 1986	By May 6, 1986	
^{133}Xe	5	45	100
^{85m}Kr	0.15	—	100
^{95}Kr	—	0.9	10
^{131}I	4.5	7.3	20
^{132}Tc	4.0	1.3	15
^{134}Cs	0.15	0.5	10
^{137}Cs	0.3	1.0	13
^{99}Mo	0.45	3.0	2.3
^{95}Zr	0.45	3.8	3.2
^{103}Ru	0.6	3.2	2.9
^{106}Ru	0.2	1.6	2.9
^{140}Be	0.5	4.3	5.6
^{141}Ce	0.4	2.8	2.3
^{144}Ce	0.45	2.4	2.8
^{89}Sr	0.25	2.2	4.0
^{90}Sr	0.15	0.22	4.0
^{238}Pu	0.1×10^{-3}	0.8×10^{-3}	3.0
^{239}Pu	0.1×10^{-3}	0.7×10^{-3}	3.0
^{240}Pu	0.2×10^{-3}	1.1×10^{-3}	3.0
^{241}Pu	0.02	0.14	3.0
^{242}Pu	0.3×10^{-6}	2.1×10^{-6}	3.0
^{242}Cm	0.3×10^{-2}	2.1×10^{-2}	3.0
^{234}Np	2.7	1.2	3.2
Approx. total	20	81	

by emitting more than 350 000 t annually [127]. This lead is either deposited directly onto the soil or it is washed down with the rain. The topsoil therefore contains much higher concentrations of lead than the deeper horizons. Concentrations of up to 100 ppm were reported from soils close to highways with heavy traffic [68].

Fortunately lead becomes fixed to the soil constituents rapidly so that penetration is not deep yet [128]. The fact that much of the lead used for manufacturing lead-alkyls is taken from deposits with a characteristic lead-isotope pattern allows one, in many cases, to distinguish between geogenic lead and anthropogenic contamination [8]. A large isotopic lead experiment has been conducted in Italy to show the emitting sources and the effect on human beings [129].

Lead also contaminates the complete hydrosphere and recent sediments [130]. Together with lead, bromine and chlorine are emitted from the exhausts of automobiles for these constituents are added to fuels to improve the removal of lead from engine and exhaust [131].

V. Lehmden [132] found that gasoline contained 10–340 ppb of mercury. Cadmium and Zinc are used as filler and stabilizer materials in tire production. During use, these elements are rubbed off and are mechanically distributed over the adjacent soils [133].

Table 15. Radioactive isotopes measured in Germany after the Chernobyl event

	Physical half-life	Principle Modes and energy of decay β-Energies (MeV)		Energies (KeV)		
^{85}Kr	10.70 years	0.672				
^{89}Sr	50.5 d	1.49				
^{90}Sr	28.8 years	0.54				
^{95}Zr	64.0 d	0.4	0.9	508	1148	
^{95}Nb	35.15 d	0.2		766		
^{99}Mo	66.0 h	1.2		740		
^{103}Ru	39.35 d	0.2	0.7	497	610	
^{106}Ru	368 d	0.04				
^{110m}Ag	250.4 d	0.09	1.5	658	885	
^{125}Sb	2.77 years	0.3	0.6	428	601	636
^{127}Sb	3.85 d	0.9	1.5	686	473	784
^{129m}Te	69.6 min	1.5		28	460	487
J	8.04 d	0.34	0.61	364	637	284
Te	78.0 h	0.2		228	50	
^{133}X	5.29 d	0.3		81		
^{133}J	20.8 h	1.2	1.5	530	875	
^{134}Cs	2.06 years	0.089	0.66	0.605	0.795	
^{136}Cs	13.0 d	0.3	0.7	819	1048	
^{137}Cs	30.2 years	0.51	1.18	0.66		
^{140}Ba	12.79 d	0.5	1.0	537	163	305
^{140}La	40.2 h	1.4	2.2	1596	487	816
^{141}Ce	32.51 d	0.4	0.6	145		
^{144}Ce	284.8 d	0.3		134	80	

Nothing has been reported so far on molybdenum-lubricants and their contamination.

Road side soils are not only contaminated by lead but also to lesser degree by cadmium, nickel and vanadium [134–136].

Mobility of Elements in Soil

Mobility of elements in soil depends on the mineralogy, the organic substances, and the redox state. In a large research project of the German Science foundation [6] mobilities of heavy metals were investigated as a function of pH: Cd, Zn, Mn, Cu and Pb are better dissolved from the soil matrix the lower the pH. At a given acid pH, Cd dissolved best. Sorption of these elements on humic acid shows high distribution coefficients in the sequence $Pb^{2+} \geq Cu^{2+} \geq Cd^{2+} \geq Zn^{2+}$.

Schemes for illuminating the bonding types of elements in soils have been developed. They take into consideration ion exchange on clays, sorption on humic substances, adsorption on iron- and manganese oxides, incorporation in carbonates and silicates. Although not all aspects can be explained by this method of step-wise (fractionated) dissolution, many principal processes can be explained by these methods.

Thornton [137] reported increasing solubility of selenium and molybdenum from Irish and British soils which have a higher pH ($\geq 6, 5$). Hornburg [138] investigated mobility and plant-availability for Cd, Zn, Mn, Pb, and Cu in soils. He applied various complexing agents to different soils, and investigated the Ca^{++} exchangeable fraction.

Hornburg and Brümmer [139] showed a connection between the Cd-extractable by EDTA and DTPA to the total Cd content in soil and to the Cd uptake by plants. Special relationships between the Tl content in soils and uptake in wheat plants was observed by Maisenbacher [140]. Only the thallium content decreases from root to grain in each nodium by a factor of 0.5. This observation could not be made for any other element. The regulations concerning metal concentrations in agricultural soils in Germany give thresholds for several elements which are surpassed by soil containing a Jurassic bituminous shale. Nevertheless there have been no observations of animal or human diseases resulting from this although As, Cd, Mo, Ni, Sb, Se, Tl, V levels are very high.

Extraction experiments with fresh samples of this soil brought low values for dissolved substances. When the highly carboniferous sediment was heated to 500 °C much higher quantities of these elements could be leached.

After heating to 1000 °C some elements were volatilized, others were transformed into other minerals which became either practically insoluble (U) or whose solubility increased (As; Cd; Ba; Cu; Cr; Li; Mo etc.)

These investigations by Puchelt and Nöltner [141] demonstrated the importance of the bonding type.

References

1. Warren HV (1972) Responsibilities of geologists in a modern world, 24th Internat Geol Congress Montreal, Symposium I, p 74
2. Dunhan K (1972) The influence of crustal resources, 24th Internat Geol Congress Montreal, Symposium I, p 16
3. Lüttig G (1972) The role of geoscience in one of the great indust nations, 24th Internat Geol Congress Montreal, Symposium I, p 41
4. Connor JJ, Feder GL, Erdman JA, Tidball RR (1972) Envir Chem in Missouri, 24th Internat Geol Congress Montreal, Symposium I, p 7
5. Campbell I (1972) Geology, environment, and citizenship, 24th Internat Geol Congress Montreal, Symposium I, p 1
6. Schwertmann U (1985) Mitteilung XIV der Geowissenschaftliche Gemeinschaftsforschung, Verlag Chemie, Weinheim, p 73
7. Walk H (1982) PhD thesis, University of Karlsruhe, FRG
8. Puchelt, H (1990) Bleiuntersuchungen von Böden an Waldstandorten in SW Deutschland, ORD Nr. 20–89.12, Landesanstalt für Umweltschutz
9. Murozumi M et al. (1969) Geochimica et Cosmochimica Acta 33: 1247
10. Wedepohl KH (1991) The composition of the upper Earth's crust and the natural cycles of selected metals. In: Merian E (ed) Metals and their compounds in the environment. Verlag Chemie, Weinheim

11. Seim R, Tisehendorf G (1990) Grundlagen des Geochemie, VEB Deutscher Verlag für Grundstoffindustrie, Leipzig
12. Rösler HJ, Lange H (1972) Geochemical Tables. Elsevier, Amsterdam
13. Paul EA, Huang PM (1980) Chem aspects of soil. In: Hutzinger O (ed) Handbook of Environ Chem, Vol 1, part A, p 69
14. Berry LG, Mason B (1952) Mineralogy, W.H. Freeman, San Franciso, chap 5
15. Taylor SR (1964) Geochimica et Cosmochimica Acta 28: 1273
16. Krauskopf KB (1989) Introduction to geochemistry, McGraw-Hill, Singapore
17. Heier KS, Billings GK (1970) Chapter 3 Lithium sections B, C, E–O, In: Wedepohl KH (ed) Handbook of Geochemistry Vol. II B1, C1, E1, F1, G1, H1, I1, K1, M, N1
18. Harder H (1974) Chapter 5 Boron sections B to 0, In: Wedepohl KH (ed) Handbook of Geochemistry vol. II
19. Rose AW, Hawkes HE, Webb JS (1979) Geochemistry in mineral exploration, Academic, New York
20. Faure G (1977) Principles for isotope geology, Wiley, New York
21. Gulson BI (1986) Lead isotopes in mineral exploration. Development in Econom Geol 23, Elsevier, Amsterdam
22. Bielicke KH, Tischendorf G (1991) Contributions to mineralogy and petrology 106, Springer, Berlin Heidelberg New York, p 440
23. Doe BR (1970) Inorganic materials 3, Springer, Berlin Heidelberg New York
24. Bowen R (1988) Isotopes in the earth sciences, Elsevier, Amsterdam
25. de Paolo DJ (1988) Nd isotope geochemistry, Springer, Berlin Heidelberg New York
26. Lerman A, Meybeck (1988) NATO ASI series, Vol 251. Kluwer, Dordrecht
27. Schachtschabel P, Blume HP, Hartge KH, Schwertmann U (1989) Lehrbuch der Bodenkunde, 12th edn, Enke, Stuttgart
28. FAO-UNESO (1974, 1988) Map of the world, vol 1. Legend, Paris
29. Tavernia R, Louis A (eds) (1982) Soil map of the European Community. Comm Eur Com, Luxemburg
30. Bowen HJM (1966) Trace elements in biochemistry. Academic, New York
31. Andrews-Jones DA (1968) Mineral Ind. Bull 11: 31
32. Seto P, Deangelis P (1978) Proceedings of the sludge utilization and disposal conference, Ottawa, Canada, p 138
33. Brüne H (1985) VDLUFA Schriftenreihe 16, Kongreßband, p 85
34. Müller G, Haamann L, Kubat R, Noe K (1987) Heidelberg Geowiss Abhandlungen, Band 13 (1987), 346 pgs
35. Essington ME, Mathgod SV (1990) Soil Sci Soc Am J 54: 385
36. Schöttle M, Wolf D, Turiam G (1990) Landesanstalt für Umweltschutz, Baden Württemberg, Sachstandsbericht 4
37. Kramar U, Puchelt H (1981) J Geochemical Explor 15: 597
38. Kramar U (1989) Geochemische Kartierung von Böden in der Umgebung des Meggener Lagers
39. Van den Boom G, Pöppelbaum, M (1980) Determination of the Presence of a Primary Mercury Dispersion around the Volcanogenic Sulphide Ore Deposit "Woodlawn", New South Wales, Australia in Geol. Jahrbuch, D 37, 15–27
40. Fletcher WK, Hoffman SJ, Mehrtens MB, Sinclair AJ, Thomson I (1987) Reviews in Econom. Geol. Vol 3, p 1
41. Arbeitsgemeinschaft Bodensystematik (1985) MiH Dtsch Bodenledl Ges, p 44
42. Beus AA, Grigorian SV (1977) Geochemical exploration methods for min. deposits, Applied, Wilmette Il, p 287
43. Boyle RW (1974) Geol Survey Can, paper 77–45, p 40
44. Stanton RE (1976) Analytical Methods for Use in geochemical Exploration, Halsted Press, New York
45. Fletcher WK (1981) Analytical Methods in geochemical prospecting, In: Govett CJS (ed), Handbook of Exploration, Geochemistry, Vol. 1, Elsevier, Amsterdam, Oxford, New York
46. Schmitt HW, Sticher H (1991) Heavy metal compounds in the soil, In: Merian E (ed) "Metals and their compounds in the environment." Verlag Chemie Weinheim, Bergstr, 311
47. Underwood EJ (1977) Trace elements in human and animal nutrition. Academic, New York, p 459
48. Goldschmidt VM (1937) The principles of distribution of chemical elements in minerals and rocks. I Chem Soc, Part 1 (1937), 655

49. Walk H (1982) should read entspr (ret 7)
50. Jenny H (1941) Factors of soil Formation, McGraw Hill, New York
51. Thorp J (1931) The effect of vegetation and climate upon soil profiles in northern and northwestern Wyoming, Soil Sci, 283
52. Kloke A (1980) Jnintelligiblo to non-German speaker
53. Leitraad Bodemsanering (1988) Netherlands Ministry for Housing, Town and County Planning, and the Environment*
54. VSBO (1986) Verordnung über Schadstoffe im Boden (= Ordinance on soil pollutants*) Hamburg Building Authorities
55. Hamburger Baubehörde (1985) Methods of evaluating the potential dangers of groundwater pollution*
56. California Assessment Manual Standards (1984) California, USA
57. Kelly RT (1979) Proceedings of the Reclaim 79 Conference, paper B2, Eastbourne, Soc Chem Ind, London
58. Ewers U (1991) In: Merian E (ed) Metals and their compounds in the environment, VCH, Weinheim, FRG, p 687
59. Zielhuis RL (1991) In: Merian E (ed) Metals and their compounds in the environment, VCH, Weinheim, FRG, p 651
60. Landesanstalt für Umweltschutz, Baden Württemberg (1989) Grenzwerte u., Richtwerte für die Umweltmedia Luft, Wasser,Boden
61. Herms U (1989) In: Behrens D, Wiesner J (eds) Beurteilung von Schwermetall kontaminationen in Boden, Dechema-Fachgespräche-Umweltschutz, p 189
62. Eikmann T, Kloke A (1991) In: Rosenkranz D, Einsele G, Harreß H-M (eds) Handbuch Bodenschutz 3590, Schmidt, Berlin, p 1
63. National Priorities List Fact Book (1986) US Env. Protection Agency, Washington DC, p 94
64. Wilmoth RC, Hubbard SJ, Burckle JO, Martin JF (1991) In: Merian E (ed) Metals and their compounds in the environment, VCH, Weinheim, FRG, p 19
65. Schalich J, Schneider FK, Stadler G (1986) Fortschr Geol Rheinld u west f 34: 11
66. Hildebrandt L (1985) 2000 Jahre Bergbau in Wiesloch, Weise, Oberanger, Munich, p 13
67. Hildebrandt L, Mohr H (1985) 2000 Jahr Bergbau in Wiesloch, Weise, Oberanger, Munich, p 15
68. Puchelt H, Schmitz-Hartmann W, Kramar U (1992) Applied Geochem: (accepted for publication)
69. Schmitz-Hartmann W (1988) Master's thesis, Karlsruhe University, FRG
70. Manz M (1990) Master's thesis, Karlsruhe University, FRG
71. Ritter J (1991) Master's thesis, Karlsruhe University, FRG
72. Rüde T (1991) Master's thesis, Karlsruhe University, FRG
73. *Schröder AV, Reuss C (1883) Die Beschädigung der Vegetation durch Rauch und die Oberharzer Rauchhüttenschäden, p 51, Berlin
74. Ernst WHO, Joosse van Damme ENG (1983) Unweltbelastung durch Mineralstoffe, Fischer, Stuttgart, p 234
75. Buchauer MJ (1973) Env Sci Techn 7: 131
76. MAGS: Ministerium für Arbeit, Gesundheit und Soziales (1977) Nordrhein-Westfalen, Düsseldorf
77. Freedman B, Hutchinson TC (1981) Source of metal and elemental contamination of terrestrial environment. In: Lepp NW (ed) Effect of heavy metal pollution in plants, Applied Science, London, p 35
78. Valkovic V (1983) Trace elements in coal, Vol II, CRC, Press, Boca Raton, Florida, p 281
79. Valkovic V (1983) Trace elements in coal, Vol I, CRC, Press In, Boca Raton, Florida, p 219
80. Kautz K (1984) In: Fortschritte der Mineralogie 62, Vol 62, Schweizerbart'sche, Stuttgart, p 51
81. Heinrichs H, Brumsack HJ (1984) Anerkennung to Ref [82] in the same volume
82. Heinrichs H, Brumsack HJ, Lange H (1984) Emissionen von Stein- und Braunkohlekraftwerken der Bundesrepublik Deutschland, Fortschritte Miner. 62, 79
83. Erzinger J, Puchelt H (1982) Erzmetall 35: 173
84. Puchelt H, Walk H (1980) Naturwissenschaften 67: 90
85. Puchelt H, Walk H (1981) Forschungsprojekt Nr 54 des Ministeriums für Ernährung, Landwirtschaft, Umwelt und Forsten Baden Württemberg

*Aus Ernst, WHO, Joosse van Damme "Umweltbelastung durch Mineralstoffe" 1983 G. Fischer Verlag Stuttgart

86. Schoer J (1984) Thallium, In: Hutzinger O (ed) The Handbook of Environmental Chemistry, vol. 3, Part C, Springer Verlag, Berlin, Heidelberg, New York, Tokyo, 143–214
87. Lehn H, Schoer J (1985) "Thallium Transfer from soils to plants: relations between chemical forms and plant uptake", In: Lekkas TD (ed) Heavy Metals in the environment, International Confernce Athens, vol. II, CEP consultants Ltd. Edinburgh, 286–288
88. Bambauer HU, Schäfer H (1984) Fortschritte der Mineralogie 62: 33
89. Schoer J, Nagel U (1980) Naturwissenschaften 67: 261
90. Fleischer M, Robinson WO (1963) Some problems on the geochem of fluorine. In: Shaw DM (ed) Studies in Analytic Geochemistry, University of Toronto Press
91. Beising R, Kirsch H (1974) VGB Kraftwerkstechnik 54: 268
92. Wenzel KF (1965) Staub 25: 121
93. Vogel J, Ottow JCG (1991) Fluoride accumulation in different earthworm species near an industrial emission source in southern Germany Bull Environ Contam Toxicol, 47, 515
94. Dässler HG (1976) Einfluß von Luftwerunreinigungen auf die Vegetation, G Fischer, Jena, GDR
95. Hluchan E, Mayer J, Abel E (1964) Pol'honospodarst, vol 10, p 257
96. Schmitt H, Moser E (1965) Erzbergbau Metallhüttenwesen, vol 18, p 111
97. Hölte W (1962) Bericht Landesanstalt Bodennutzungschutz, Nordrhein-Westfalen, p 43
98. Lovelace CJ, Miller GW, Welkie GW (1968) Atmos Env 2: 187
99. Oelschläger W (1971) Fluoride Quart Rep 4: 80
100. McNeal JM, Rose AW (1974) Geochim Cosmochim, Acta 38, 1759
101. Saager R (1984) "Mercury" in Metal, Resources and Applications Bank von Tobel, Zürich, 99
102. Kaiser G, Tölg G (1980) "Mercury" in Hutzinger (ed) The Handbook of Environmental Chemistry, Vol. 3, Part A, 1
103. Weiss HV, Koide MV, Goldberg ED (1971) Mercury in a Greenland ice sheet: Evidence of recent input by man. Science, 174 (11), 692–694
104. Hentschel Th, Priester M (1990) Quecksilberbelastungen in Entwicklungsländern durch Goldamalgation im Kleinbergbau und aufbereitungstechnische Alternativen, Erzmetall 43, 331
105. Whitby LM, MacLean AJ, Schnitzer M, Gaynor JD (1977) Sources storage and transport of heavy metals in agricultural watersheds. International Reference Group on Great Lakes Pollution from Land Use Activities. International Joint Commission, Windsor, Ontario
106. Webber J (1981) Trace elements in agriculture. In: Lepp NW (ed) Effect of heavy metal pollution on plants, Vol 3, Applied Science, London, p 159
107. Williams CH, David DJ, Aust J (1977) Soil Res 15, 59–68
108. Furrer OJ (1980) Landwirtschaftlicher Wert des Klärschlamms, EAS-Seminar, Basel, p 24
109. Umweltbundesamt (Federal Ministry for the Environment) (1988/89) Luftquditätskriterien fur Blei, p 128
110. Förstner U, Stiefel R (1978) Chem Zeitung 102: 161
111. Schriftenreihe Umweltschutz, BUS (1984) Nr 26 Kompostierung
112. Hermite PL, Ott H, (1984) Proc. of 3rd Intenat Symp on Processing and Use of Sewage Sludge, Reidel, Dordrecht
113. Sinclair WA, Stone EL, Scheer CF Jr (1975) Hort Sci 10: 35
114. Miles JRW (1968) J Agric Food Chem 16: 620
115. Bishop RF, Chesholm D (1962) Canad J Soil Sci 42: 77
116. Delas J, Dartigues A (1970) Aun agronom (Paris) 21: 603
117. Acimovic M (1974) Bilt Hmelj Siral 6: 11
118. Nagai T, Koga H (1975) Tottori Daigaku Nogakubu Hokoku 27: 34
119. van Rhee JA (1969) Rigks Fac Landbouw Gent 34: 682
120. Eversmeyer MG, Browder LE (1974) Plant Dis Reg 58: 469
121. Brown BD, Rolston DE (1980) Transport and transformation of methyl bromide in soils. Soil Sci 130, 68–75
122. Gowen JA, Wlersma OB, Thai H (1973) Pesticid Monit J 10: 111
123. Poelstra PM, Frissel J, van der Klugt N, Bannink JW (1973) Neth J Agric Sci 21: 77
124. Koelzer W (1988) Lexicon der Kernergie. Kernforschungszentrum, Karlsruhe, p 207
125. Eisenbud M (1987) Environmental radioactivity, Academic, Orlando
126. Gumbrecht D, Heller H, Kindt A (1987) Auswirkung des Reaktorunfalles in Tschernobyl auf die BRD, Veröffentlichungen der Strahlungschutzkommission, Bd 7, Gustav Fischer, Stuttgart, BRD
127. Newland LW, Daum KA (1982) Lead. In: Hutzinger O (ed) The Handbook of Environmental Chemistry, Vol 3, Part B, 1–26

128. Dörr H, Münnich KO, Mangini A, Schmitz W (1990) Gasoline Lead in West German Soils, Naturwissenschaften 77, Springer Verlag, 428–430
129. Colombo A, Facchetti S (1988) The Isotopic Lead Experiment: Impact of petrol lead an human blood and air. Final report: Commission of the European Communities, EUR 12002
130. Ter Haar GL (1986) Fate of gasoline lead in the Environment. In: Stokes P (ed) Pathways, Cycling and Transformation of Lead in the Environment, the Royal Society of Canada, The Commission on lead in the environment
131. Ewers U, Schlipkötter NW (1991) Lecid in "Metals and their Compounds in the Environment," Merian E (ed) Verlag Chemie Weinheim, New York, Basel, Cambridge 970–1014
132. Lehmden DJV, Jungers RH, Lee RE (1973) Evaluation of analytical techniques for Determination of trace elements in coal, by fly ash, fuel oil, and gasoline. Abstract of Papers American Chemical Society 165, Abs pap ACS, 32, 73 MNOR
133. Kötter L, Niklauß M, Toennes A (1989) Erfassung möglicher Bodenverunreinigungen auf Altstandorten, Kommunalverband Ruhrgebiet, Arbeitshefte Ruhrgebiet A 039 S. 238–247
134. Lagerwerff JV, Specht AW (1970) Environ Sci Technol 4: 83
135. Hutchinson TC (1972) University of Toronto Publ EH 2: 27
136. Ward NJ, Brooks RR, Roberts E, Boswell CR (1977) Environ Sci Technol 11: 917
137. Thornton I (1981) In: Lepp NW (ed) Effect of heavy metal pollution on plants, Vol 2, Applied Science, London, p 1–84
138. Hornburg V (1991) Bonner Bodenkundl Abhandlung 2: 228
139. Hornburg V, Brümmer GW (1990) VDLUFA 32, Kongreßband, p 821
140. Maisenbacher P (1991) Ph.D thesis, Karlsruhe University, FRG
141. Puchelt H, Nöltner T (1990) Interactions of naturally occurring aqueous solutions with the Lower Toarcian oil shale of S. Germany. In: Heling D, Rothe P, Förstner U, Stoffers P (eds) Sediments and environmental chemistry, Springer, Berlin Heidelberg New York, p 291

Evolution of Matter and Energy

M. Taube

CH-8956 Killwangen, Switzerland

Nuclear Primordial Evolution of Matter 68
  Nature, Laws and Constants................................. 68
    The Theory of Everything 68
    Space, Time, Symmetries, Natural Constants................... 68
  Elementary Forces and Elementary Particles 68
    Elementary Forces and Particles 68
    Four Elementary Forces 70
    Elementary Stable Particles: Quarks and Electrons............. 70
    Interaction of Elementary Particles and Forces............... 70
    Binding Energy of Nuclides............................... 75
    Mutual Connection of Matter and Energy 77
  Atomic Nuclei 77
    Nucleosynthesis in the Early Stages of the Universe 77
    Primordial Synthesis of Hydrogen and Helium............... 78
  Stars as Nuclear Reactors 80
    Galaxies, the Largest Cosmic Structures 80
    Stars as Nuclear Energy Sources and the Synthesis of Elements.... 80
    Synthesis of $C\ N\ O$ in Stars............................. 82
    Synthesis of Metals in Stars.............................. 83
    Synthesis of Heavy Elements in Supernovae 84
    Neutron Stars, Black Holes and Nucleosynthesis 85
    Synthesis of Light Elements due to Cosmic Rays 86
    Stability of the Natural Existing Elements................... 86
    Magic Nuclear and Electronic Numbers 87
    Abundance and Electronegativity of Elements................. 87
  Evolution of the Sun...................................... 88
    Origin of the Protosun, Energy Flow and Chemical Composition .. 88
    Sun Today as Energy Source............................... 89
    Future Evolution of the Sun: Red Giant 91
  Solar Abundance of Elements 91
    Primordial Abundance of Elements......................... 91
  The Distant Future of Nuclear Matter 94
    Cycling of Cosmic Matter 94
    The Very Distant Future: Black Holes, Neutrinos, Photons 96
  Nuclear Boundary Conditions for the Existence of the Realm of
  Chemical Phenomena 97

The Handbook of Environmental Chemistry
Volume 1 Part F, Ed. O. Hutzinger
© Springer-Verlag Berlin Heidelberg 1992

Evolution of Terrestrial Matter and Energy   97
    Chemical Reactions in Interstellar Space.   97
        Distribution of Cosmic Matter   97
        Gas-Dust Clouds, Chemistry, Energy Sources.   98
        Comets and Their Chemistry.   99
        Meteorites and Cosmic Dust   102
    Protoplanets and the Evolution of Planets   104
        Origin of Solar Planets   104
        Origin of Protoplanetary Clouds.   104
        Planets, Chemical Composition.   105
    The Planet Earth.   105
        Terrestrial Abundance of Elements   105
        The Stability of Terrestrial Matter.   108
        Chemical Changes in the Past   109
    Terrestrial Flow of Energy and Matter.   110
        Energy Flow: in the Past, Today and in the Future   110
    Abiogenic Chemical Evolution of Terrestrial Matter.   111
        The Conditions of Early Chemical Evolution   111
        Primordial Soup; Abiogenic Synthesis.   112
        Chemical Impact of Comets and Meteorites on the Terrestrial
            Environment   113
        Evolution of the Terrestrial Hydrosphere in the Past, Today and
            in the Future   114
        Evolution of the Terrestrial Lithosphere in the Past, Today and
            in the Future   114
        Water as Erosion Factor in the Lithosphere   114
Terrestrial and Extraterrestrial Life.   115
    General Definition of Life   115
        Life; Coupling of Matter and Energy.   115
        Life: The Need for a Definition   116
        Spontaneity and Universality of Life   118
        High Order (Low Entropy) and Life   119
        Life and the Inflow of Free Energy   119
        Life and the Regulation of the Flow of Matter and Energy   119
        Reproducibility; Stability and Changeability   120
    Material and Energetical Carriers of Living Systems   121
        General Remarks About the Carrier of a Living System.   121
        Which Elementary Forces Could be Appropriate?.   121
        Why Chemical Forces?   121
        Why is Hydrogen one of the Carriers of Life?.   122
        Why is Oxygen the Second Elementary Carrier of Life?   123
        Why is Carbon the Third Life-Carrying Element?.   124
        Why is Nitrogen the Fourth Elementary Carrier of Life?   125
        Phosphorus or Another Element as the Fifth Carrier of Life?.   125
        Other Elements in Living Systems.   126
        Thermal Conditions for the Existence of Life.   127

Biosphere in the Past... 128
   The First Living Structures: Protobionts and Eobionts........... 128
   Environment and the Biosphere in the Past 129
   Further Evolution of the Biosphere.......................... 130
   Extinctions and Discoveries................................ 131
   Catastrophes Caused by Terrestrial Events.................... 132
The Biosphere Today and in the Future.......................... 133
   Biomass and the Production at Present....................... 133
   Direct and Indirect Use of Biomass by Humans................ 135
   Productivity of the Marine Biosphere 137
   Marine Food Production; Marineculture?..................... 138
   Present and Future Productivity of the Continental Biosphere..... 140
Biosphere and the Use of Solar Energy........................... 141
   The Efficiency of Green Plants 141
   Number of Species—Past, Present and Future 142
   Biogenic Limits of Food Production? 143
The Distant Future of the Terrestrial Biosphere 144
Man and the Flow of Energy and Matter: Past, Present, Future 145
Man and the Biosphere: Past, Present, Future 145
   Global Human Population 145
   Mankind in the Distant Future............................. 146
   How Much Surface for Mankind?........................... 146
   Another Solution: Extraterrestrial Colonisation. 149
Man and the Flow of Matter: Past, Present, Future............... 150
   Total Flow of Matter..................................... 150
   The Best is Taken Now................................... 153
   Quasi-Irreversible Matter Flow: Concrete, Metals.............. 155
   Irreversible Material Flow: Fossil Fuels 155
   Reversible Flow of Matter: Water, Food, etc.................. 156
   Matter Recycling and Energy Use........................... 157
   Global Flow of Materials in the Future 158
Food for Everyone 159
   How Much Food Does a Man Need?........................ 159
   Food Production and the Use of Technological Energy 161
   Continental Food Production: Agriculture 161
   Industrial Synthetic Human Food........................... 162
Global Climate: Past, Present, Future........................... 163
   Global Climate Machine 163
Sources of Free Energy on the Earth 164
   Needs for Technological Energy: Past, Present, Future........... 164
   How Much Energy Does a Man Need?....................... 166
   Non-Renewable Resources for Energy: Past, Present, Future 169
   Renewable Energy Resources.............................. 173
   Energy Perspectives for the Very Distant Future................ 174
   Global Thermal Pollution 175
Very Distant Future of Mankind: Mega and Giga Years 176

Distant Future of Mankind: Large Scale Factors   176
Stability of Universe, Galaxies and the Sun.   177
Stability of Planet Earth   177
Extraterrestrial Connections   178
Abbreviations, Symbols.   179
References   180

Nuclear Primordial Evolution of Matter

Nature, Laws and Constants

The Theory of Everything

Maybe during our time, at the end of the twentieth century or at the beginning of
the next century, the old dream of natural sciences will become reality—the theory
which unifies all phenomena, the natural and the artificial, the smallest and the
largest, the simplest and the most complex, the past and the future, the here and the
elsewhere. The possibility of the emergence of the theory of everything seems not
to be as remote as before, even if some theoretical physicists are too optimistic. Not
only on the grounds of pure symmetry, the number of sceptical theoreticians who
do not believe such "theories of everything" equals the number of enthusiasts. The
"theory of everything", TOE, is probably approaching. There exist different ways
of achieving the TOE. One of them, which seems to have good prospects is the
theory of "superstrings". This theory proposes that all particles—real particles, the
'bricks of the Nature' (fermions), and also field particles, the carriers of
forces,being 'the mortar of nature' (bosons)— are not pointlike but are stringlike.
The superstrings theory not only describes all known forces (including gravitation), but gives an intimate coupling between forces, particles, space and time
[1, 2, 3].

Space, Time, Symmetries, Natural Constants

The real "theory of everything" must explain why the Universe relies on three-
dimensional space and one-dimensional, unidirectional time, why natural phenomena are of high symmetry and why the symmetry is so often broken, why the
natural constants, such as light velocity in vacuum, the gravitational constant,
Planck's constant and Einstein's cosmological constant, have one and not another
value and how they evolve during time or why they are time independent [4].

Elementary Forces and Elementary Particles

Elementary Forces and Particles

Almost all present physical theories claim that the Universe has an origin, the so-
called 'Big Bang'. This event lies around 15 gigayears ago. How can we define the
boundary conditions of the emerging Universe? There exists a hypothesis that the
boundary condition of the Universe is that it has no boundary and therefore no

boundary conditions. Did the Universe emerge from nothing? Where did the energy come from to create matter? The answer, according to recent theories, is that the energy was borrowed from the gravitational energy of the vacuum. The Universe has an enormous debt of negative gravitational energy, which exactly balances the positive energy content of matter [5].

For an extremely short time after the origin, around 10^{-44} s, the Universe contained only one elementary force, which we could call the 'Superunified Force', and only one class of elementary particles, which we could call the 'Superunified Particle'. The temperature at this moment would be very high, around 10^{32} K, with a density equal to 10^{94} kg/cub.m. The properties of space-time and matter-energy under these conditions are far from our experimental possibilities and are assumed on the basis of emerging physical theory, the so called 'Theory of Everything'. The most important feature is the fact that a number of consequences of the process evolving during the Big Bang could be predicted and checked by observed conditions in the present state of the Universe.

As time passes the Universe increases in volume and its temperature drops. This dramatic phase of the evolution of the Universe is called 'the inflation phase' and the appropriate model that of the 'inflationary Universe'. This model is very successful in explaining a number of rather unique properties of the Universe, but it cannot be accepted without some reservation [6, 7, 8].

After around 10^{-36} s the temperature falls to 10^{28} K and the density decreases dramatically. Under these conditions the Superunified Force spontaneously divides into two forces: Gravitational and Grand-unified Force. Also the superheavy Superparticles decay into two kinds of matter—particles and antiparticles. The number of particles (common matter) and antiparticles (antimatter) is equal.

After further evolution up to 10^{-24} s the dramatic inflationary phase finishes, the temperature falls to 10^{24} K and the density decreases greatly. At this moment the Grand-unified Force divides into two forces—a Strong Force and a Unified Force—and particles undergo further evolution. The conditions which exist in this moment are the subject of a rather well-developed physical theory, the so-called Grand-Unification Theory (GUT). One of the simplest forms of this theory predicts an observable phenomenon, the decay of protons, with a half-life of around 10^{33} years and the creation of massless neutrinos. However, experiments in this field up to the present time have not yielded positive results [9].

The expansion of the Universe and the evolution of the existing forces and particles develop further. Because the Unified Force does not preserve the symmetry between particles and antiparticles, the spontaneous decay of both kinds of particles (matter and antimatter) produces more particles than antiparticles. The particles and antiparticles interact, resulting in annihilation. In this process both types of matter disappear and photons, neutrinos and other particles emerge. These particles are:

—fermions, that is the very bricks of the Universe, such as quarks, electrons and neutrinos
—bosons, (field particles), that is the mortar acting between the bricks, such as massless gravitons (the carriers of the Gravitational Force), massless gluons (the

carriers of the Strong Force) and relatively heavy W- and Z-bosons (the carriers of the Unified Force).

When the age of the Universe equals 10^{-12} s, the temperature falls to 10^{16} K and two significant phenomena occur. The Unified Force splits into a Weak Force and an Electromagnetic Force. The carrier of the latter is the massless photon. The quarks begin to fuse into three-quark agglomerates: protons and neutrons. The previously created electrons move independently from the protons.

Four Elementary Forces

When the environment has a temperature lower than one Petakelvin (10^{15} K), which corresponds to a kinetic energy of around 100 GeV (gigaelectronvolt) four elementary forces exist. These forces allow one to explain all phenomena, without exception, beginning from nuclear fission and fusion, through the realm of galaxies, stars and planets to the realm of living beings and information processing in the brain of an intelligent observer [10, 11].

It is of importance to know that among all four elementary forces (mentioned in order of increasing strength), gravitational, weak, electromagnetic and strong, only the electromagnetic force is both attractive and repulsive. The other three elementary forces act only attractively. A material structure resulting from the action of attraction forces only, achieves the rather trivial form of a more or less homogeneous sphere. Examples of these are: a spherical star resulting from gravitational attraction and the more or less spherical atomic nuclei resulting mostly from the attraction of the Strong Force. A more sophisticated structure, for example a molecule results from the Electromagnetic Force, which contains both attraction and repulsion. Also, a living being, not only here and now on the Earth but elsewhere in the Universe and at all times, relies on the electromagnetic force and on electrically charged stable particles, protons and electrons (Fig. 1, Table 1, 2).

Elementary Stable Particles: Quarks and Electrons

In spite of the enormous development of physics the definition of the term 'elementary' is not clear enough. Figure 2 gives a very rough, speculative and far from being accepted, but quite illustrative, scheme of the mutual connection of the elementary particles (Fig. 2, Table 3).

Interaction of Elementary Particles and Forces

The mutual interaction of elementary forces and elementary particles and mediators is rather complex, but could be schematically illustrated by the following table.

From the large number of different particles, only a small number are involved in the specific manifestation of the electromagnetic force, which is the chemical force. From the point of view of chemical reaction it is clear that particles must be appropriately stable, because a chemical reaction needs a time of order of magnitude significantly larger than 10^{-9} s in which to take place. The only stable

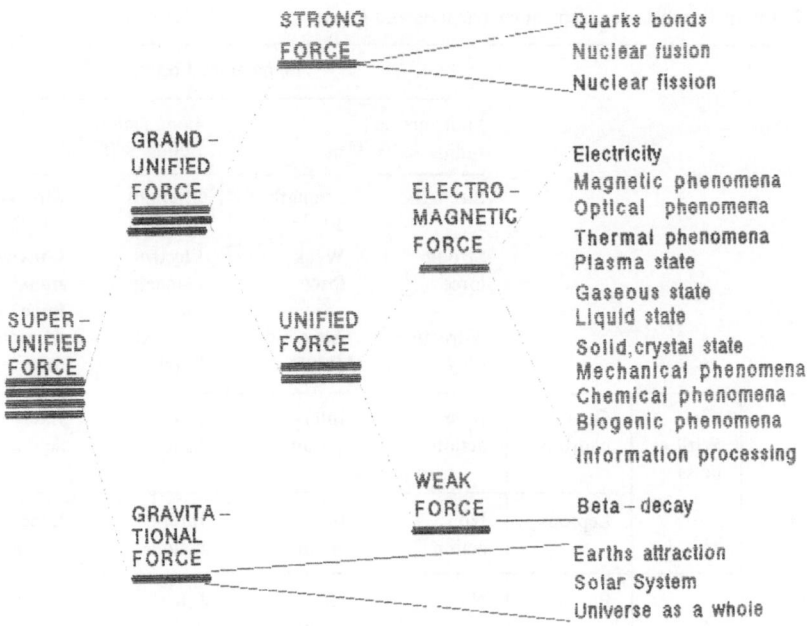

Fig. 1. Elementary forces are mutually connected

Table 1. Attractive and repulsive action of elementary forces. Note: In parentheses is given the strength of the elementary force expressed in dimensionless units, the so-called coupling constant.

		Radius of Action	
		Very short radius $< 10^{-15}$ m	Very long radius $> 10^{26}$ m
Force	Very large	Strong Force (around 1) Attraction only	Electromagnetic Force (1/137) Attraction and Repulsion
Strength	Very small	Weak Force (10^{-12}) Attraction only	Gravitational Force (10^{-38}) Attraction only

particles are the proton, bounded neutron (so-called nucleons) and electron. All other approximately 100 hadrons and leptons are very unstable, with a half-life shorter than 10^{-9} s. Here and elsewhere, the actors in the play of chemical reactions are only these stable particles mentioned.

Table 2. Interaction between elementary particles and forces

			Elementary Forces			
			Short action radius $< 10^{-15}$ m		Long action radius $> 10^{26}$ m	
			Strength 1	Strength 10^{-12}	Strength 1/137	Strength 10^{-39}
			Strong force	Weak force	Electro-magnetic force	Gravita-tional force
			Attraction only	Attraction only	Attraction Repulsion	Attraction only
Elementary Particles	With mass	Quarks hadrons	Inter-action	Inter-action	Inter-action	Inter-action
		Leptons	No action	Inter-action	Inter-action	Inter-action
	Mass less	Photon	No action	No action	Inter-action	Inter-action
		Graviton	No action	No action	No action	Inter-action

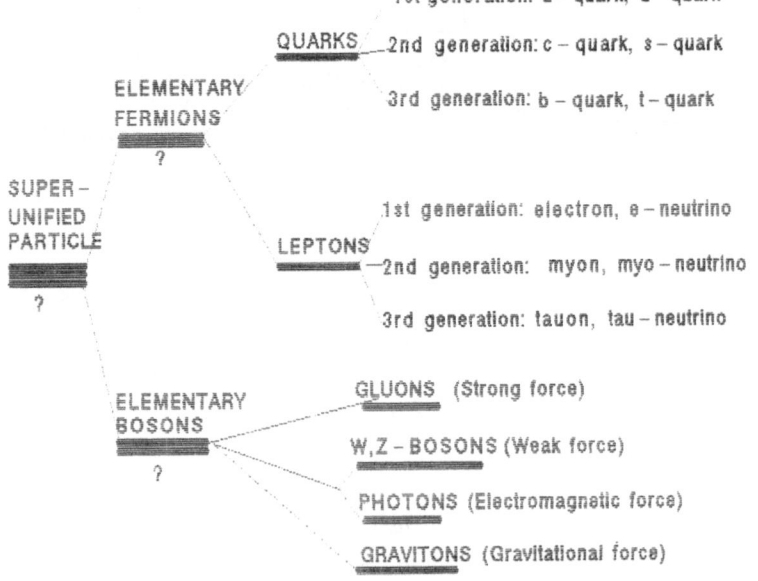

Fig. 2. Mutual connection of elementary particles

Table 3. Elementary mediators and particles

Properties of particles	"Mortar" (Mediators) Bosons				"Bricks" (Particles) Fermions	
	Many particles in single state				Only one particle in single state	
Important examples	Photon	Gluons	Graviton	Bosons	Leptons	Quarks
Symbol	gamma			W^+, W^- .Z^o	e^-, e^+	q
Electrical charge	0	0	0	+1, 0, −1	+1, 0, −1	+1/3, +2/3 −1/3, −2/3
Mass	0	0	0	85 GeV	>0	≫0
"Ordinary matter" (first generation)					Electron	Up-quark
					Electro neutrino	Down-quark
"Strange Matter" (second generation)					Muon	Strange-quark
					Muo-neutrino	Charme-quark
"High Energy Matter" (third generation)					Tauon	Top-quark
					Tau-neutrino	Bottom-quark

Let us now consider quarks, electrons and neutrinos as elementary 'bricks' and to try to explain other phenomena. At first let us try to analyse some exotic properties of quarks. We will limit ourselves to a discussion concerning only 'ordinary' light quarks, being members of the 'first generation' of elementary particles (Tables 3, 4).

Table 4. Quarks and leptons of the first generation

Particle	Mass (MeV/c²)	Electric charge	Baryon number	Spin h/2π	Elementary force
Quarks					
Up-quark	7	+2/3	1/3	1/2	Strong and ElMag
Down-quark	5	−1/3	1/3	1/2	Strong and ElMag
Leptons					
Electron, e	0.511	−1	0	1/2	Weak and ElMag
e-Neutrino	0.00001ᵃ	0	0	1/2	Weak only

ElMag = Electro-Magnetic Force
ᵃ still not directly proven

The question of the possible existence of an internal structure of quarks is now discussed, but the problem of 'elementary particles' has been shifted to the much deeper level of 'string' particles. Quarks have some peculiarities: they cannot be directly observed and their electrical charge equals 1/3 or 2/3 of the elementary charge of an electron. These rather strange properties are additional argument for the 'elementariness' of quarks.

The case of light particles, the leptons, is simpler. Leptons (e.g. electrons and neutrinos) show no internal structure. From this point of view they are elementary particles, (see Table 3) and their electrical charge equals always $+1, 0$ or -1. Some questions are of importance. For example, why does a 'down' quark (d), with an electrical charge of $-1/3$ have a larger mass then the 'up' quark (u) with electrical charge $+2/3$? This results in a neutron having a greater mass than a proton. Where e.c. = electrical charge and the mass, m in megaelectronvolts/c^2:

1 u-quark $+2$ d-quark $=1$ neutron (m $=939.552$)
(e.c. $= +2/3$) $+2$ (e.c. $= -1/3$) $=$ (e.c. $=0$)

2 u-quark $+1$ d-quark $=1$ proton (m $=938.259$)
2 (e.c. $= +2/3$) $+$ (e.c. $= -1/3$) $=$ (e.c. $= +1$)

The result is that the neutron is slightly heavier than the proton, with a mass difference equal to m $= 1.293$ MeV/c^2, which corresponds to a difference of only 0.13 percent. This mass difference allows beta-minus decay, which is controlled by the Weak Force (mass given in MeV/c^2):

$$\text{neutron} = \text{proton} + \text{electron} + \text{e-neutrino} + \gamma\text{-photon}$$
$$(939.552) = (938.259) + (0.511) + (0.000012) + (0.782)$$

The peculiarity of the mass difference between the 'down' quark and the 'up' quark results in a sequence of events which allows the emergence of the chemical realm as we know it. The neutron is spontaneously unstable against the transformation. into a proton; the proton cannot undergo this kind of transformation. The nucleons which were produced in a short period during the Big Bang as a mixture of equal number of protons and neutrons, were transformed into free protons and atomic nuclei such as deuterons, helium-3, helium-4, etc. If the neutron is lighter than the proton, then the Universe should contain, in terms of atomic weight around 93 per cent of stable neutrons and the rest of heavy matter, in the form of helium and other elements. In such a case only very small amounts of hydrogen, in the form of deuterium, would be possible. The evolution of cosmic matter without a large amount of hydrogen could not open the way to 'organic chemistry'. The second decisive effect would be the impossibility of forming long-lived stars (see 'main sequence' stars) because deuterium burns so readily in a star's interior that the stars would explode instead of burning for gigayears, as is the case. These conditions would result in the failure of intelligent life to emerge.

Some questions are of utmost importance from the point of view of the stability of the Universe, and especially from the point of view of the stability of atoms and molecules. The stability of atoms relies on the simple fact that the electrical charges of the proton and the electron are exactly equal, though of opposite sign. The measured absolute values of the positive and negative charges are equal to within an accuracy of around 10^{-21}. We know that the proton is an agglomerate of three quarks and the electron is indivisible, but we do not yet know why 2 up-quarks, each with electrical charge of $+2/3$, and 1 down-quark with electrical charge of $-1/3$ are together electrically absolutely equal to the electrical charge of 1 electron (Fig. 3).

Today we know very little, if nothing, about the origin of the different masses of different particles, in particular and of highest importance, of the origin of heavy particles, quarks and the light leptons, e.g. electrons. The mass of a proton is around 1836 times greater than the mass of an electron. This ratio controls chemical interaction. A lighter electron or a heavier proton (and neutron) would result in an increase in the diameter of atoms and decrease in the strength of the resulting chemical forces. A heavier electron and a lighter proton would cause a decrease in the diameter of atoms and the resulting chemical forces would be increased. Both cases would prevent the stability and existence of larger molecules, especially macromolecules. The result of this would be a lack of living beings and the lack of any intelligent observer in the Universe.

Binding Energy of Nuclides

Before discussing the origin of the elements and their different isotopes, some remarks about the binding energy of atomic nuclei are required. The interaction of the components of atomic nuclei results from the attractive of the Strong Force and the repulsive action of the Electromagnetic Force. The attractive force acts

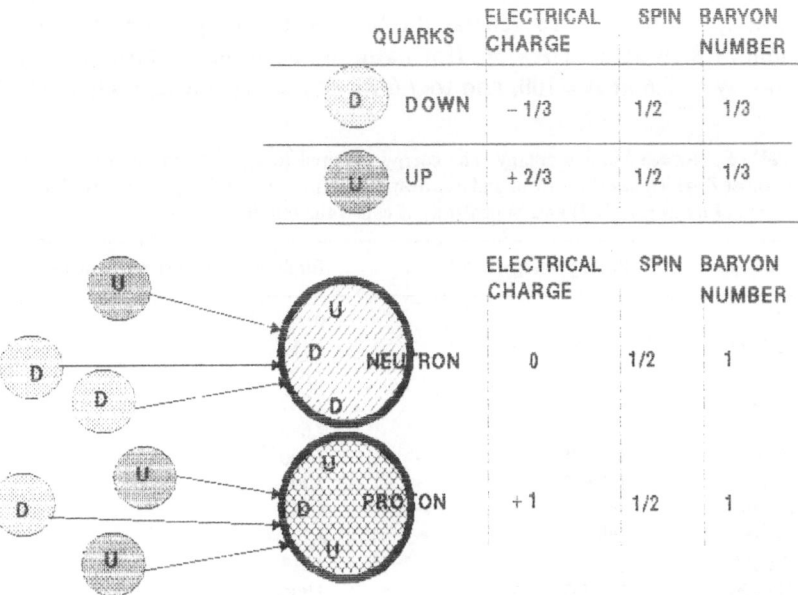

Fig. 3. Synthesis of protons and neutrons from quarks

between protons and neutrons in exactly equal strength. The repulsive action of the Electromagnetic Force, however, acts only between electrically positively charged protons. This is the reason for the rather complex relationships in the binding energy which exist in the interior of atomic nuclei, described in simple terms as follows.

The fusion of two protons is impossible because the repulsive force is larger than the force of attraction.

The fusion of a proton with a neutron leads to the synthesis of a stable deuteron (the atomic nucleus of a heavy hydrogen atom—deuterium) and the release of 2.224 megaelectronvolts (MeV) of energy that is around 1.112 MeV per nucleon. For the sake of comparison, 1 electronvolt per molecule corresponds to 96.5 kilojoules per mole. The binding energy of helium-4 equals 28.26 MeV, that is, 7.05 MeV per nucleon . The atomic nuclei of the light elements Li, Be and B have lower binding energies. The next atomic nuclei C-12 is more strongly bonded, with a binding energy equal to 7.68 MeV per nucleon (Table 5).

The binding energy per nucleon shows some small peaks at so-called 'magic nuclear numbers'. These magic nuclear numbers are 2, 8, 20, 28, 50, 82 and 126. The existence of the exceptionally stable magic nuclides results from the nature of the mutual interaction of the attraction of the strong (nuclear) force and the repulsion of the electromagnetic force acting between the electrically charged protons. It is clear that the most stable nuclides are at the same time the most abundant. Especially important, from the point of view of the realm of chemical interaction, is that helium-4 is a very stable nuclide. It is also of greatest importance from the point of view of chemical interactions that two other elements, carbon and oxygen, have the most abundant isotopes, in the form of C-12 and O-16, which are the product of the fusion of three or four atomic nuclei of helium-4, respectively. In general the binding energy per nucleon rises to a maximum of 8.79 MeV at $A = 56$, (for naturally occuring nuclides, Fe-56), then drops slowly to 8.6 at $A = 100$, and to 7.6 for the heavy nuclei, such as U-238.

Table 5. Nuclear binding energy. The energy required to separate an atom of atomic number Z, which has N neutrons and a corresponding mass number A $(A = Z + N)$, into Z atoms of hydrogen (H-1) and N neutrons, is called the binding energy.

Nucleus	Binding energy/MeV	Binding energy per nucleon/MeV
H-1	0	0
H-2 (D)	2.224	1.112
H-3 (T)	8.481	2.827
He-4	28.296	7.047
C-12	92.162	7.680
C-14	105.285	7.520
O-16	127.620	7.796
Ca-40	342.056	8.551
Fe-56	509.952	8.792
Ru-100	861.939	8.619
Hg-200	1581.209	7.906
U-235	1783.890	7.591

The differences in the binding energies of atomic nuclei allow the release of free energy due to fusion of the light elements, e.g. the fusion of D and T producing He-4 in a thermonuclear controlled power reactor, and the fission of very heavy atomic nuclei, e.g. U-235 and Pu-239, producing the 'fission products' in fission power reactors. (see Fig. 4).

Mutual Connection of Matter and Energy

The intimate relation between matter and energy is the basis for the whole spectrum of phenomena in Nature. Table 6 illustrates some of these relationships beginning at black holes, through chemical forces to thermal energy (Table 6).

Atomic Nuclei

Nucleosynthesis in the Early Stages of the Universe

Only when the age of the Universe equalled 10^{-4} s, with a temperature of 10^{12} K and density 10^{17} kg/cub.m would neutrons and protons begin to fuse due to the nuclear force, which is a kind of Van der Waals attraction for the Strong Force [14].

When the Universe was only 0,1 ms old it would still be relatively hot, with a temperature of about one Terakelvin. The numbers of protons and neutrons are approximately equal, but the free neutrons are unstable and decay, according to recent data, with a half-life of 607 s. Under these conditions, protons fuse with

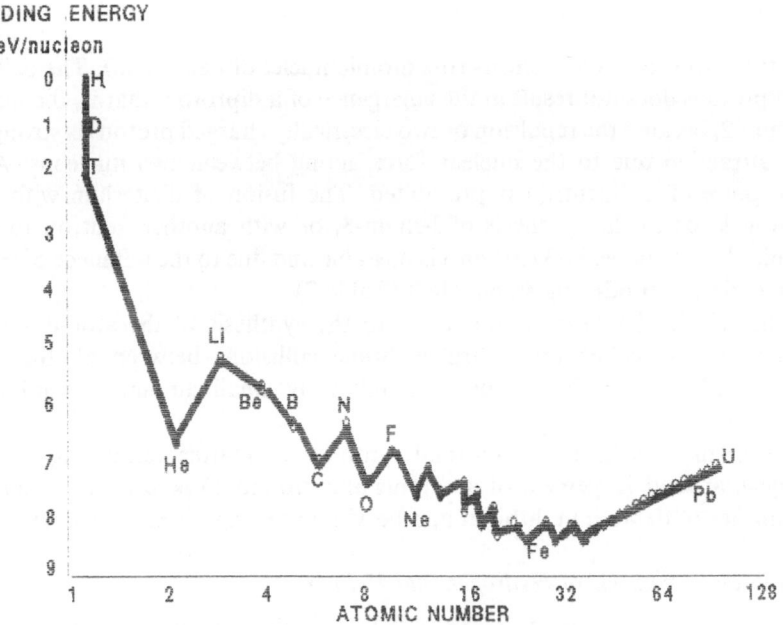

Fig. 4. Binding energy of atomic nuclei

Table 6. Mutual coupling of matter and energy

Relative	Energy per nucleon	Strong force	Electro-magnetic	Weak force	Gravitation
1	1 GeV				
		Annihilation of antimatter			Collapse of two black holes.
10^{-2}	10 MeV				Matter
		Fusion of light atomic nuclei			falling into black
		Fission of heavy atomic nuclei			hole
		Alpha decay			
10^{-4}	100 keV			Beta decay	
10^{-6}	1 keV				
10^{-8}	10 eV				
			Chemical energy		
			Solar radiation		
			'Biological energy'		
			Chemical energy		
10^{-10}	0.1 eV		Van der Waals' energy		
			Information		
					Free fall
			Therm. energy 300 K		on Earth
10^{-12}	0.001 eV				

neutrons to produce deuterons (the atomic nuclei of deuterium). The collision of two protons does not result in the emergence of a diproton, that is, the nucleus of helium-2, because the repulsion of two electrically charged protons is stronger than the attraction due to the nuclear force acting between two nucleons. Also the emergence of a dineutron is prohibited. The fusion of deuterium with another proton leads to the synthesis of helium-3, or with another neutron to tritium. While He-3 is stable, H-3 (tritium) is unstable and due to the influence of the Weak Force, decays producing stable He-3 (Table 7).

The fusion of two deuterons leads to the synthesis of the atomic nucleus of helium-4, also called an α-particle. Some collisions between all the particles mentioned lead to the synthesis of relatively small amounts of lithium and beryllium.

From this results the primordial synthesis of matter containing mostly hydrogen, around 75 percent of all atoms and around 25 percent of helium-4. The quantities of deuterium, lithium and beryllium are significantly smaller.

Primordial Synthesis of Hydrogen and Helium

A vast source of free nuclear energy is locked up in the fusion of hydrogen H-1, resulting in the synthesis of He-4 (and two positrons and two neutrinos). The

Table 7. Stable elementary particles as components of atoms

			Mortars Photons Gravitons
		Bricks	
Almost all other particles	Free neutron		Proton[a] Neutron in nuclei Electron
10^{-15}	1	10^{15} Half-life in seconds	10^{30}

[a] According to the Grand Unified Theory, a proton decays with a half-life of around 10^{32} years

amount of energy equals 26.7 MeV per atom of helium-4, which corresponds to 2.574 Peta joule/kg. The direct fusion of 4 atoms has an extremely low probability. Fusion occurs in stars as a result of the so-called proton-proton chain (see Fig. 5) or by means of 'catalytic' synthesis due to the CNO (carbon-nitrogen-oxygen) cycle (see Fig. 7). In a fusion reactor, that is in a controlled thermonuclear reactor, the synthesis of He-4 relies on the reaction between deuterium(D) and tritium(T):

$$D + T = He\text{-}4 + Neutron + 17.6 \text{ MeV}$$

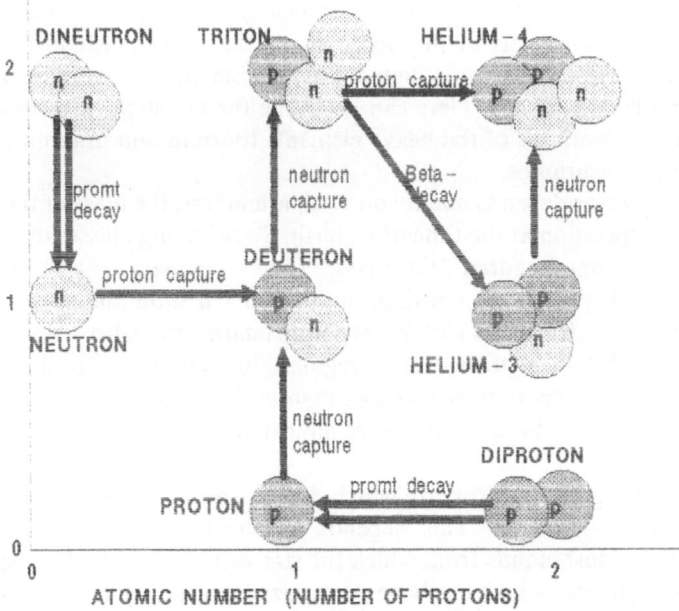

Fig. 5. Primordial synthesis of deuterium and helium

The released neutron is used for the synthesis of tritium following its reaction with lithium-6 (the so-called 'fusion breeding process):

$$\text{Neutron} + \text{Li-6} = \text{He-4} + \text{T} + 4.8 \text{ MeV}$$

Stars as Nuclear Reactors

Galaxies, the Largest Cosmic Structures

The Universe contains probably 100 giga galaxies, with the average galaxy having 100 giga stars. The average mass of a galaxy is roughly equal to 10^{41} kg and the age of the galaxies is probably not much shorter than that of the whole Universe, that is around 15 gigayears. There are many different types of galaxies. Our galaxy, called the Milky Way, is a typical spiral. There probably exists in the centre of numerous galaxies, including our own, a central body in the form of a large black hole, of mass equal to a million solar masses.

Between one hundred to ten thousand galaxies are gravitationally bound and belong to a cluster of galaxies [14, 15, 16].

The chemical evolution of galaxies is not very well understood, because it depends on the evolution of individual stars of different masses and different ages, and also on the evolution of intergalactic non-luminous matter.

Stars as Nuclear Energy Sources and the Synthesis of Elements

The continuous expansion of the Universe results in continuous cooling and a decrease of density. Under these conditions no nuclear synthesis is possible, because the repulsion force of positively charged protons and atomic nuclei of helium-3 and helium-4 is of the order of one MeV per nucleon. However, the possibility of high temperature, high density, stable 'nuclear reactors' is offered by the internal layers of stars. Here can begin the further steps of nuclear synthesis, leading to the synthesis of the heavy elements thorium and uranium, even to the superheavy transuranids.

The life of a star depends mainly on two parameters, the mass of the star and its chemical composition at the time of its birth. Stars having masses of between one-hundredth and one hundred solar masses (one solar mass = $1.99 \cdot 10^{30}$ kg) and a luminosity of between one ten-thousandth and ten thousand solar luminosities (one solar luminosity = $3.90 \cdot 10^{26}$ W) generate temperatures between one thousand and 100 megakelvin in their central regions. In some short-lived stars, such as supernovae, the temperatures reach as high as 5 gigakelvin. Here occur nuclear processes which lead to the synthesis of almost all other elements, proceeding at a high rate.

The second significant property which decides the later development of a star is its chemical composition. This depends on the history of material from the protosolar gas-dust clouds from which the star originated. The first generation of stars includes matter which has been synthesized during the first few minutes after the Big Bang, that is mostly H, D, He-3, He-4, Li, Be and B. During the evolution of these first-generation stars matter undergoes different nuclear processes, resulting in the synthesis of other light elements with C, N, O, Ne, Mg, Si and S as the most important and most significant. A number of stars of the first generation then explode (e.g. supernovae) and dissipate stellar matter into galactic space. Under the impact of gravitation, gas-dust clouds evolve which result in the emergence of stars, mostly double or triple stars, which are stars of the second generation. The life of these stars passes differently to that of first generation stars. In some of these stars are produced metals which include the heavy metals, such as thorium and uranium. After the explosion phase of these stars matter forms new gas-dust clouds, from which emerge the third generation of stars, which from the very beginning includes all stable nuclides. The Sun is a star of the third generation, and therefore includes all elements including uranium. The history of the Earth depends very strongly upon this, that the protosolar cloud was significantly enriched in stellar matter originating in third-generation stars [17, 18, 19].

Stars of all three generations and of different composition evolve in general over a relatively short time, of the order of magnitude of less than one million years, from the protostar phase up to the so-called 'main sequence stars'. The 'main sequence stars' live from between hundreds of millions to ten or more gigayears, or even longer (Fig. 6).

The later lives of the stars depends very much on their masses. Very lightweight stars, with masses lower than 0.07 solar mass, are too small to achieve in their centres, during the gravitational contraction phase, the appropriate temperatures and pressures which allow nuclear reactions to begin. These stars live a very long time as 'brown dwarfs' or even 'black dwarfs'. It seems that the number of these dwarfs is relatively high and their total mass is greater as the mass of all the luminous stars combined.

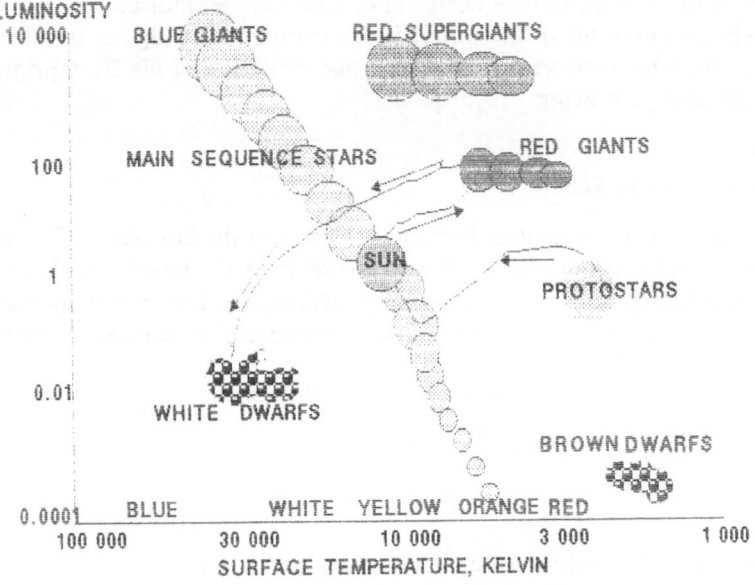

Fig. 6. Russel-Hertzsprung star diagram

Lightweight stars, having a mass of 0.07 to about 4 solar masses, go through the hydrogen-burning phase and then through the helium-4 burning phase. In the beginning, the small amount of Li, Be and B burns, because the initiation temperature is lower than 10 megakelvin. The phase of hydrogen burning, both the proton-proton process or catalytic hydrogen burning (CNO-cycle), lasts a very long time and corresponds to 'main sequence' stars. After approximately a tenth of the hydrogen has been burnt a new phase begins. Our central star, the Sun, belongs to this class and is a typical 'main sequence' star. More about the evolution of the Sun will be mentioned later.

In the centre of an old main sequence star the accumulated helium achieves a relatively high temperature and pressure and helium burning begins. This phase of star evolution is called the 'red giant' phase. The outer layers of the star expand, the stellar radius increases dramatically and the surface temperature decreases to around one thousand degrees Kelvin. This phase cannot last longer than tens of millions of years. The next phase is a contraction to a small but relatively dense star, the so-called 'white dwarf'. In these stars all nuclear reactions cease and the star cools slowly over tens of gigayears.

Middleweight stars, with 4 to 8 solar masses, go through an explosive phase of 'carbon burning' resulting in the synthesis of Mg and its neighbouring elements. Heavyweight stars, of 8 to 60 solar masses, go through the 'supernova' phase (see Fig. 8). This type of stars could be observed on 27th January 1987 both optically and by the measurement of neutrinos generated by it.

Here is now the correct place to make some remarks about the role which a main sequence star plays as the source of free energy for chemical reactions on neighbouring planets. The stars surface has a mean temperature of 5000 to 7000 K and is a source of photons with energies of 1 to 4 electronvolts. The chemical force corresponds to the same level of energy (1 electronvolt per molecule equals 96 500 J/mol). The period of life of these stars is of the order of 10 gigayears, which is a period of time which allows the emergence and evolution of life from primordial soup to an intelligent being.

Synthesis of CNO in Stars

He-4 is a product of the nuclear burning of D during the hot phase after the Big Bang; the so-called primordial synthesis. However in the interior of dense large stars, of the red giant type, He-4 burns producing C-12 in a 'helium burning' process, in this peculiar case also called the '3α process' (α-particle = He-4):

$$He\text{-}4 + He\text{-}4 \rightarrow Be\text{-}8 \text{ (unstable)}$$

$$Be\text{-}8 + He\text{-}4 \rightarrow C\text{-}12$$

$$\overline{\qquad\qquad\qquad\qquad\qquad\qquad}$$

$$3\alpha \rightarrow C\text{-}12 + 7.24 \text{ MeV}$$

The next step of helium burning is the synthesis of O-16:

$$C\text{-}12 + He\text{-}4 \rightarrow O\text{-}16 + 4.73 \text{ MeV}$$

The relation between the atomic nuclei of He-4, Be-8 and C-12 is an example of the very subtle and very important interaction of the elementary forces. Fig. 4 shows the nuclear binding energy of these nuclei, from which it must be clear that the instability of Be-8 is the reason for the very large stability and cosmic abundance of C-12.

If Be-8 were more stable it would also be more abundant, even more abundant that carbon or oxygen. In such a situation the Universe would be occupied by BeO and there would not be enough oxygen to form H_2O. The consequences for the emergence of life are self-evident (Table 8).

A further step in the He-burning process leads to the synthesis of Ne-20, Mg-24, Si-28, S-32, Ar-36 and even Ca-40; heavier products being of smaller significance. The temperature for this synthesis equals between 1 and 2 gigakelvin.

The synthesis of N-14 also occurs within the interiors of red giants due to the fusion of C-12 with 2 protons and one beta-decay. There are some peculiarities in this process in that the nuclide N-14, having an odd number of protons 7 and also an odd number of neutrons 7, is so abundant. All other odd-nuclei have extremely low cosmic abundance (Fig. 7).

Synthesis of Metals in Stars

The synthesis of metals up to $Z=26$, that is Fe, from the point of view of thermodynamics, occurs with the release of free energy. The synthesis of elements with atomic number higher than $Z=26$, that is from Fe up to Th and U, is only possible due to the input of free energy from outside. The most abundant source of free energy is the gravitational contraction of a star. In extreme cases the contraction of the internal layers of a massive star results in the explosion of the outer layers. This is the case in a supernova explosion.

During a slower contraction a number of atomic nuclei which are rich in neutrons, such as C-13, split into a neutron and C-12. The dual occurrence of high temperature and free neutrons leads to the synthesis of a large number of isotopes of medium mass nuclides, up to the isotopes Pb-208 ($Z=82$) and Bi-209 ($Z=83$).

Table 8. Origin of the isotopes of C, N, O

Isotope	Relative Abundance (atomic) in the Sun	Synthesis process		
		Helium-burining	Equilibr. CNO	Hot CNO
C-12	1500	X		
C-13	17		X	
N-14	480		X	
N-15	1.8			X
O-16	2800	X		
O-17	1.0		X	X
O-18	5.5	X		

The elements with atomic number $83 < Z < 89$ are short lived, and therefore the synthesis occurring during a slow contraction cannot produce a significant proportion of the heavier elements [18, 19].

Synthesis of Heavy Elements in Supernovae

During the explosion of a supernova, the extremely high temperature of 5 gigakelvin and large flux of free neutrons which originate from the disintegration of iron nuclei into protons and neutrons allow the synthesis of isotopes of Pb, up to Pb-238 or even heavier isotopes with an extremely large surplus of neutrons. The very heavy isotopes are short-lived and decay under the action of the Weak Force. This decay process, transforming the neutrons into protons with the emission of electrons and neutrinos, leads to the synthesis of unstable but very long-lived isotopes of U-238, U-235 and Th-232. In our cosmic neighbourhood, the supernova explosion probably occurred more than 5 gigayears ago—just before the emergence of the Solar System. The half-lives of these isotopes of U and Th are

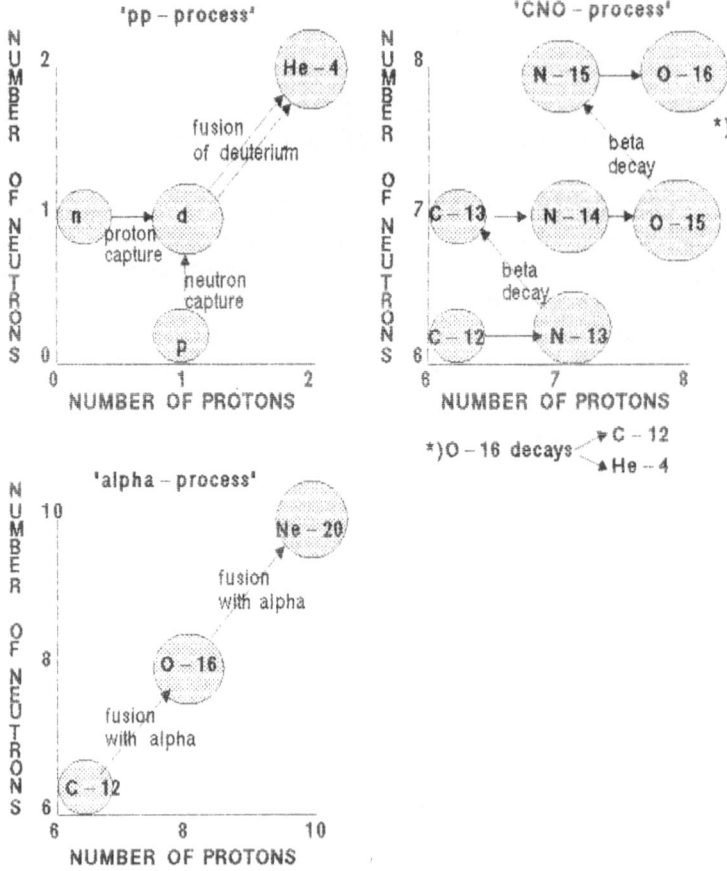

Fig. 7. Synthesis of carbon and oxygen; helium burning

some gigayears. Therefore, they continue to exist on the Earth today and will continue to exist for tens of gigayears [20].

On 23 February 1987 the explosion of a supernova, called SN 1987a, was observed in the Large Magellanic Cloud, a satellite galaxy of our Galaxy, the Milky Way, about 170 000 light years away. The progenitor star was the star called Sanduleak 69°202, a blue supergiant with an effective temperature of roughly 12 000 K. SN 1987a was detected visually and by the direct measurement of neutrinos. The latter results from the collapse of the core of the supernova into a 10-km diameter neutron star with a mass equal to 1.5 solar masses. Some scientists hypothesise that the neutron star has a larger mass and therefore subsequently collapses to become a black hole. The emergence of radioactive Co-56 fits in with the theoretically predicted nucleosynthesis that Ni-56 (the double magic nuclide with 28 protons and 28 neutrons) decays into Co-56 and subsequently into Fe-56, the stable nuclide. It must be assumed that significant amounts of Th and U have also been synthesized and dissipated into the surrounding interstellar space by the explosion [21, 22] (Fig. 8).

Some astrophysicists postulate that supernovae can heat comets residing in a comet cloud surrounding the Sun, far beyond the last planet. As a supernova explosion occurs in our Galaxy every 20 years, this phenomenon together with the influence of passing stars could move some comets in the direction of the Sun, that is, in the neighbourhood of the Earth. Such an indirect influence of supernovae on our planet could be of importance.

Neutron Stars, Black Holes and Nucleosynthesis

The evolution of a heavy star reaches its conclusion in its explosion as a supernova, though not the whole mass of the star is ejected into space. A part, around one tenth, remains in the centre (but not less than 1.5 solar masses) and is compressed by the explosion shock wave into an extremely dense star, a neutron star with a density equalling the density of atomic nuclei—10^{17} kg/cub.m. Only a thin surface

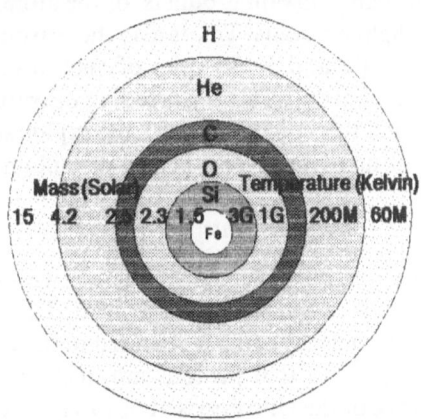

Fig. 8. Scheme of supernova II

layer exists, containing some protons and electrons, resulting from the equilibrium state: neutron $\longleftrightarrow$ proton + electron. Some investigators have likened a neutron star to a giant nucleus with atomic mass of about 10^{57}. There are some observable neutron stars, among them the so-called 'pulsars'. There exists the possibility that some nucleosynthesis processes are induced on the surface of a neutron star [23].

If the mass of a neutron star increases, for example due to the infall of matter from outside, the gravitational force causes a further collapse, the radius of the star decreases slightly and a new type of cosmic body emerges—the black hole. At present time no such stellar object has been observed, 'but some objects are suspected of being black holes. In the direct neigbourhood of a black hole strong gravitational acceleration are experienced by other stellar objects (dust-gas clouds) passing nearby and it is possible that these effects could be observed from Earth. The presence of very massive black holes, of the order of magnitude of tens of millions of solar masses, has been suspected in the centre of numerous galaxies, even in our own Galaxy, and also in quasars. Some investigators have claimed that the accretion disk of supermassive black holes could convert hydrogen into helium and produce small amounts of heavier elements. In spite of these rather exotic ideas, from the point of view of nucleosynthesis black holes seem to be a graveyard for matter and not a source of it.

Synthesis of Light Elements Due to Cosmic Rays

The light metals Li, Be and B have a rather complicated origin. In spite of their synthesis during the hot phase of the Universe, just after the Big Bang, the atomic nuclei of these elements, all having binding energies lower than He or C, are good nuclear fuels. During the formation of protostars, when temperature is only some megakelvin, all these elements undergo nuclear burning and are lost. The total amount of these elements on our planet cannot thus have a primordial origin. They cannot also be synthesized in the interior of stars. Their probable origin is the impact of cosmic rays (atomic nuclei having very large kinetic energy and coming from a large distance) on atomic nuclei of C, O, N and other elements. The results of such collisions are probably 'fission products' of the atomic nuclei being struck, the atomic nuclei of the light elements. This is also the reason for the extremely low cosmic abundance of Li, Be and B. In spite of such a complicated origin the amounts of lithium on the Earth enable the breeding of tritium in a fusion reactor in such large quantities that this could guarantee the exploitation of thermonuclear reactors for the whole of mankind over a period of millions of years [24].

Stability of the Natural Existing Elements

The number of naturally occurring elements, together with man-made heavy and superheavy elements, equals at the present time around 110. The obvious, stable isotopes are so stable that no decay can be measured in them. However, the Grand Unified Theory (GUT) assumes that protons are unstable, having a half-life of around 10^{33} years. The age of the Universe is about 15 gigayears and the best guess as to the length of life of the 'stable' nuclides must correspond to the half-life of the proton, according to the GUT. All isotopes of some of medium mass elements, Tc and Pm, are rather short-lived, and the naturally occurring products of the spontaneous decay of Th-232, U-235 and U-238 are very short-lived. All nuclides with $Z > 92$ are relative short-lived, and earlier hopes of synthesising elements with $Z = 114$ or $Z = 118$, which according to some models of the atomic nucleus should be long-lived, that is with half-life of the order of gigayears, must be abandoned (Fig. 9).

Magic Nuclear and Electronic Numbers

Outside the hot interior of stars, where the temperature is lower than 1500 K, and in particular lower than 400 K, the electromagnetic force acts on atomic nuclei and electrons in both attractive and repulsive ways, and produces atoms and molecules. The scene for chemical interaction is then set. The numerous members of the family of the elements show some specific properties, among others the existence of the noble gases. The elements with the 'electronic magic number' (analogous to the 'nuclear magic number') do not take part in chemical interactions, at least to a first approximation (Table 9).

Abundance and Electronegativity of Elements

Chemical reactions in cosmic space depend on numerous parameters. For the sake of simplicity only two parameters are discussed here—abundance and electro-negativity. Electronegativity does not include the noble gases, which do not play a

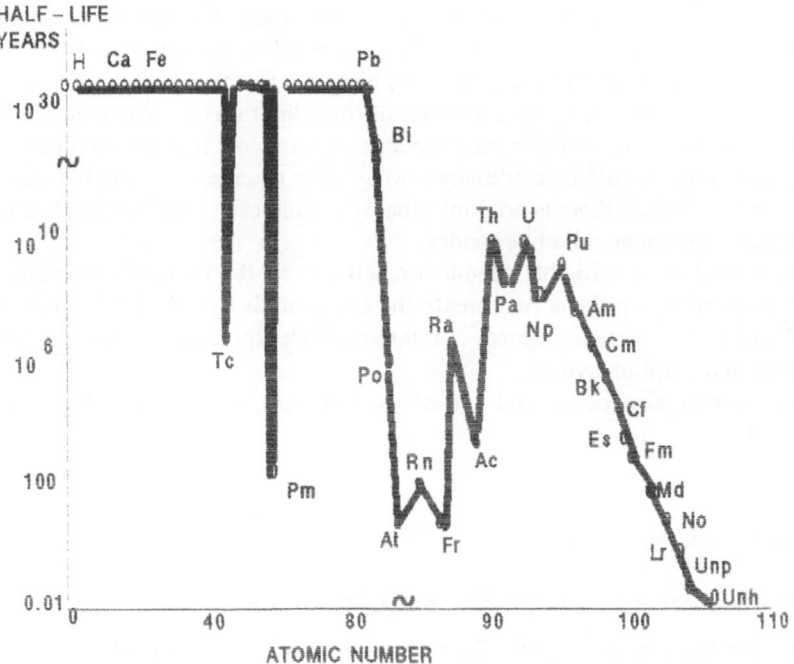

Fig. 9. Stability of elements

Table 9. Nuclear and electronic magic numbers

								Ordinal numbers								
0	5	10	15	20	25	30	35	40	45	50	55	60	65	70	75	80
							Nuclear magic number for neutrons									
2	8			20	28					50						82
							Nuclear magic number for protons									
2	8			20	28					50						82
He	O			Ca	Ni					Sn						Pb
							Electronic magic number for elements									
2		10	18				36			54						
He		Ne	Ar				Kr			Xe						

role in chemical interaction in cosmic space. The second element in the list of cosmic abundance, helium is approximately one hundred times more abundant than all other elements. This fact is of the highest significance in our discussions.

The largest difference in electronegativity occurs between fluorine and lithium or beryllium, and between oxygen and both of these light metals. The cosmic significance of these compounds is extremely small, but let us consider the question, "what could happen if.?" If the abundance of the light metals Li, Be and B were comparable to that of carbon or nitrogen, then the greatest percentage of oxygen would be bound up in the oxides of these light metals. These would be the most abundant compounds in the Universe and it seems that the chances for the emergence of life would be extremely low, and the emergence of intelligent beings almost zero. Though there is no doubt that life could exist which depends on these nonvolatile, extremely alkaline oxides.

The second most probable chemical reactions, on the basis of abundance and electronegativity, are those that create the compounds hydroxyl, OH, and water, HOH. Other hydrides also appear in relatively high abundance, such as ammonia, methane and sulphur hydride.

Some oxides also occur and are of interest: CO, CO_2, SiO, SiO_2 and FeO (Table 10).

Evolution of the Sun

Origin of the Protosun, Energy Flow and Chemical Composition

Interstellar space is not empty. On average it is filled with 10 million atoms per cubic meter at a temperature of not much higher than 3 K. The amount of matter in interstellar space is small percentage by weight of the total stellar matter. Part of

Table 10. Electronegativity of elements and its abundance. Numbers give relative cosmic abundance of elements, assuming that the abundance of silicon equals $1\,000\,000 = 1$ Mega.

					25 G			
G = Giga					H			

G = Giga
M = Mega
k = kilo

	55	0.7	6.6	7.9 M	2.1 M	17 M	710
	Li	Be	B	C	N	O	F

	57 k	1.0 M		1.0 M			5 k
	Na	Mg	Al	Si P S			Cl

3.5 k	59 k	860 k			8
K	Ca	Ti Fe	As Se	Br	

6	26		1.1
Rb	Sr	Noble metals Te	I

0.3	4.2		0
Cs	Ba	Lanthanide	At

0	0	0.03	0	
Fr	Ra	Actinide		117

0	1	2	3	4

Electronegativity in Pauling's units

this interstellar matter belongs to large gas-dust clouds, some of which are light years in extent, occupying roughly 5 percent of the galactic volume. Over a period of around 5 gigayears a gas-dust cloud with a mass of some 10^{30} kg in our part of the Galaxy (Milky Way) emerges, including the matter created in a parental second-generation star. Probably, during this time, one or even two explosions of supernovae in the neighborhood took place, which enriched the protosolar cloud with heavy elements, including gold, thorium and uranium. Approximately 5 gigayears ago a star of the third-generation with a mass of about $2 \cdot 10^{30}$ kg emerged in this cloud, the Sun (Fig. 10).

Sun Today as Energy Source

The statement that the Sun belongs to the main sequence stars gives us a good deal of very important information. Main sequence stars have the following inherent properties:

—longevity of the order of magnitude of 10 gigayears. The Sun, with a typical mass, will probably live in this phase for around 10 gigayears. But because it is

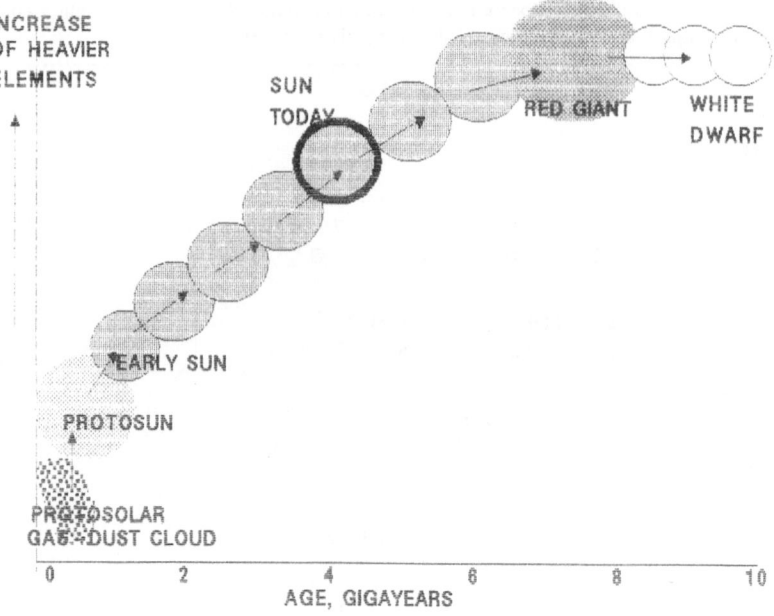

INCREASE
OF HEAVIER
ELEMENTS

SUN
TODAY

RED GIANT   WHITE
                DWARF

EARLY SUN

PROTOSUN

PROTOSOLAR
GAS-DUST CLOUD

0   2   4   6   8   10
AGE, GIGAYEARS

Fig. 10. Evolutionary path of the sun

already about 5 gigayears old it can only remain in this stage of its evolution for a further 5 gigayears. This longevity results from the rather slow burning of hydrogen (H-1) in the centre of the Sun. If the Sun contained the heavier isotope H-2 (deuterium), instead of H-1, than the rate of burnup would be much faster and a violent explosion would be unavoidable, with all the consequences for the birth of the Earth.

—longevity is an alternative expression for the absence of large catastrophic events. However, it is too much to assume that the Sun can not pass through events which are not too large, but which are intensive and which may have a dangerous impact on the Earth. The well-known cycles in the Sun of 11 and 22 years are rather small in magnitude, but not without an influence on the terrestrial climate.

—main sequence stars have surface temperatures of some thousand Kelvin. The spectrum of radiation from the surface of a main sequence star can be compared to the radiation of a black body. The maximum in the spectrum of main sequence stars lies at some electronvolts, that is in the wavelength region of 500 nanometers.

—it is obvious that the energy of photons emitted by the Sun exactly corresponds to the free energy required for numerous chemical reactions. This is the reason why main sequence stars play the role of carrier of free energy for the atoms and molecules in their neighborhood. Chemistry and the main sequence stars are intimately coupled (Table 11).

The evolutionary path of the Sun begins with the emergence of a protosolar gas-dust cloud, continues through the protosun phase to become a main sequence star. It will probably then pass through the phase of being a red giant and go on to

Table 11. Sun today [25, 10]

Parameter	Dimension	Value
Age	gigayears	4.6
Type	–	Main sequence star
Life duration in this form	gigayears	4
Star generation	–	Third
Next evolutionary type	–	Red giant
Energy source	–	Hydrogen burning
Mass	kg	$1.99 \cdot 10^{30}$
Radius	m	$6.960 \cdot 10^{8}$
Density, mean	kg/cub. m	1410
Luminosity	W	$3.86 \cdot 10^{26}$
Temperature, surface	K	5785
Specific power, mean	W/kg	0.000188
Surface energy flux	MW/m^2	63
Centre of Sun		
—density	kg/cub. m	135 000
—temperature	K	14 600 000
—pressure	megabar	200

become a white dwarf. The Sun also acts as the source of terrestrial material, including the very light and the heavy elements.

Future Evolution of the Sun: Red Giant

Today the Sun is approximately 5 gigayears old. Models of the evolution of the main sequence stars of the third generation allows a rather well-defined prediction of the further evolutionary path to be made. During the next period of more than 4 gigayears, the Sun will insignificantly contract and its luminosity increase. After the period of being the main sequence star, the Sun will pass into the phase of being a red giant. During a relatively short period of some hundreds of millions of years the outer radius of the Sun will increase by some 40 times and its surface temperature will fall to around one thousand kelvin.

Its luminosity will increase by some 360 times, and the terrestrial solar constant will increase from its present value of 1.367 kW/m^2 to around 460 kW/m^2. The effective temperature on the night side of the Earth will increase from its present value of 253 K to about 1170 K. All the oceans will have evaporated before the Sun reaches the full red giant stage. Later, the Sun will cool and be transformed into a white dwarf, with a longevity of order of magnitude of tens of gigayears.

Solar Abundance of Elements

Primordial Abundance of Elements

The chemical interaction of elements depends not only upon the temperature, pressure and quality of free energy, but also on the relative concentration of the elements. The estimation of the chemical composition of such a complex and inaccessible system as the Solar System is a rather difficult process. We know the chemical composition of the Sun's surface and of the atmosphere of the planets, and we have analysed probes returning from the surface of the Moon or sent to Mars and Venus.

Halley's comet was an excellent opportunity of analyzing a cosmic body directly, by analytical instruments passing through its coma. We know very well the chemical composition of meteors and meteorite, but knowledge of the composition of the Earth is limited to a depth of 12 km, and only one probe to this depth has so far been made. Further drillings to this depth are, however, in preparation (Table 12, Fig. 11).

Table 12. Primordial solar abundance of elements. Most probable relative abundance of the primordial solar system elements. Data are numbers of atoms normalized to an abundance of silicon $= 10^6$ [26]

Atomic Number Z	Element	Abundance
1	H	$25.0 \cdot 10^9$
2	He	$2.5 \cdot 10^9$
3	Li	55
4	Be	0.73
5	B	6.6
6	C	$7.9 \cdot 10^6$
7	N	$2.1 \cdot 10^6$
8	O	$17.0 \cdot 10^6$
9	F	710
10	Ne	$1.4 \cdot 10^6$
11	Na	$57 \cdot 10^6$
12	Mg	$1.0 \cdot 10^6$
13	Al	$80 \cdot 10^3$
14	Si	$1.00 \cdot 10^6$
15	P	$8.6 \cdot 10^3$
16	S	$480 \cdot 10^3$
17	Cl	$5.0 \cdot 10^3$
18	Ar	$220 \cdot 10^3$
19	K	$3.5 \cdot 10^3$
20	Ca	$59 \cdot 10^3$
21	Sc	35
22	Ti	$2.4 \cdot 10^3$
23	V	290
24	Cr	$13.5 \cdot 10^3$
25	Mn	$8.7 \cdot 10^3$
26	Fe	$860 \cdot 10^3$
27	Co	$2.2 \cdot 10^3$
28	Ni	$48 \cdot 10^3$
29	Cu	450
30	Zn	$1.4 \cdot 10^3$
31	Ga	44
32	Ge	113
33	As	6.5

Table 12. (contd.)

Atomic Number Z	Element	Abundance
34	Se	63
35	Br	8.0
36	Kr	25
37	Rb	6.4
38	Sr	26
39	Y	5.4
40	Zr	11
41	Nb	0.85
42	Mo	2.5
43	Tc[a]	0
44	Ru	1.8
45	Rh	0.33
46	Pd	1.32
47	Ag	0.50
48	Cd	1.30
49	In	0.17
50	Sn	2.4
51	Sb	0.27
52	Te	4.8
53	I	1.16
54	Xe	6.61
55	Cs	0.37
56	Ba	4.2
57	La	0.46
58	Ce	1.20
59	Pr	0.18
60	Nd	0.87
61	Pm[a]	0
62	Sm	0.27
63	Eu	0.10
64	Gd	0.34
65	Tb	0.06
66	Dy	0.41
67	Ho	0.09
68	Er	0.026
69	Tm	0.04
70	Yb	0.25
71	Lu	0.038
72	Hf	0.18
73	Ta	0.021
74	W	0.127
75	Re	0.052
76	Os	0.72
77	Ir	0.65
78	Pt	1.42
79	Au	0.19
80	Hg	0.40
81	Tl	0.17
82	Pb	3.1
83	Bi	0.136

Table 12. (contd.)

Atomic Number Z	Element	Abundance
84–89	Po, At, Rn, Fr, Ra, Ac[a]	0
90	Th	0.032
91	Pa[a]	0
92	U	0.009
93–>110	Np, Pu, Am, Cm, etc[a]	0

[a] Elements with no stable isotope; all isotopes are short-lived.

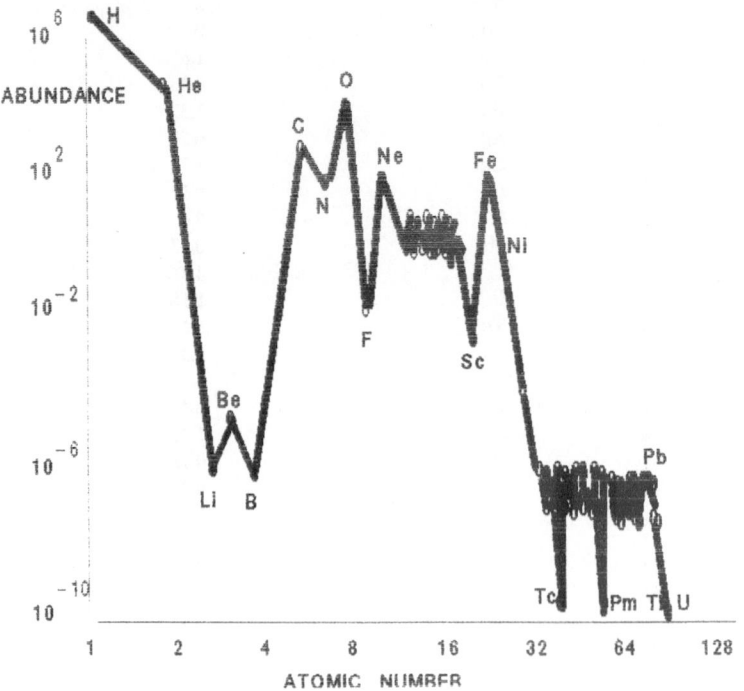

Fig. 11. Solar abundance of elements (simplified)

The Distant Future of Nuclear Matter

Cycling of Cosmic Matter

The abundance of elements in the Universe is probably not far from that in the Solar System. However, this is valid only for the present moment. In the first minutes after the Big Bang the chemical composition of the Universe was very simple and contained hydrogen, helium and small amounts of lithium and beryllium. It is self-evident that, parallel to the aging of the Universe, its chemical composition will change dramatically.

Without going into the details of this very complex series of nuclear processes, it seems to be possible to formulate some general trends over the next tens of gigayears:

—the amount of hydrogen will decrease, because the rate of its synthesis is smaller by far than the rate of its burnup and the production of helium,

—the amount of helium will probably slightly increase, because its rate of synthesis is lower than the burnup of helium in the red giant stars,

—the amount of medium mass, including iron, will increase, because these are the products of supernovae and cannot be used in future nucleosynthesis,

—the most decisive trend is doubtless the neutronisation of cosmic matter, that is the clear increase of the number of neutrons, not only in the nuclei of medium and heavy elements, but also in the form of neutron stars, which seem to have very long longevity,

—the other clear trend is the 'disappearance' of baryonic matter, that is the atomic nuclei, and a large part of leptonic matter in the form of electrons in atoms into black holes. However, it may be that matter does not disappear, but only the baryonic and leptonic properties disappear. Gravitational mass and electrical charge are fully conserved in black holes,

—another result of all these processes is the increasing number of neutrinos, which have been produced by different nuclear processes, radioactive decays, during the collapse into neutron stars and just before the gravitational collapse into black holes (Fig. 12).

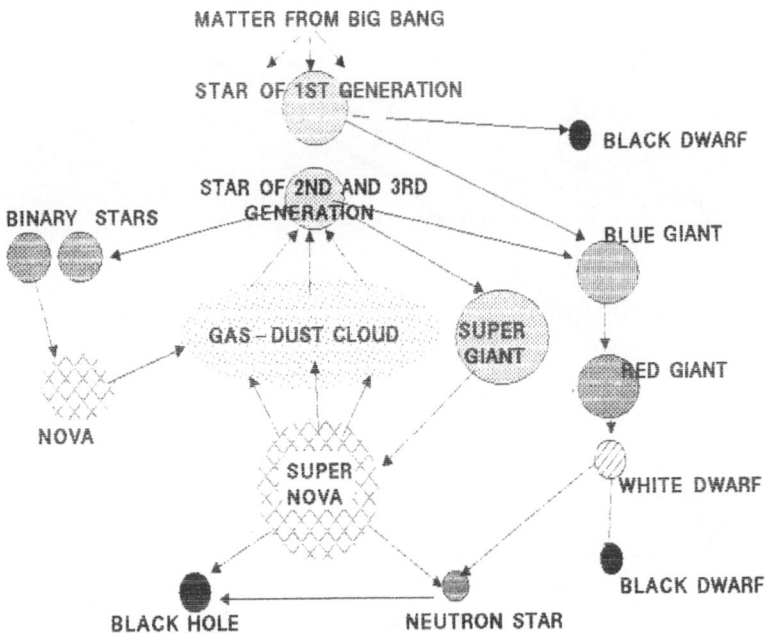

Fig. 12. Cycling of cosmic matter

The Very Distant Future: Black Holes, Neutrinos, Photons

If we wish to analyse the very distant future of matter, we must answer the question concerning the future of the Universe. Almost all theories of cosmology, the science concerning the 'beginning of everything', rely on the hypothesis of the Big Bang. This momentary 'act of creation' seems to be well understood. But what is the fate of the Universe? What will happen in a hundred gigayears, in a million gigayears, in billions of exayears and so on?

Once more without going into details, we can distinguish between two different scenarios for the Universe:

—the closed Universe; after another 45–50 gigayears of expansion, the Universe will come to a period of continuous contraction. After an additional 60–70 gigayears it will collapse to a very dense state, which can be likened to the state during the Big Bang. This end of the Universe is called the 'Big Crunch', and is the end of our story.

—the open Universe; the expansion will continue or, at least, will gradually come to an end and the Universe will exist for an infinite time. The first danger comes after a period of around 10^{33} years, when the spontaneous decay of protons will destroy 'normal matter'. In this scenario all matter disappears into black holes, with a large number of neutrinos and photons remaining which can escape being captured by black holes. If the Universe exists longer, even the black holes seem to disappear, due to the so-called 'Hawking radiation', and are transformed into photons. The Universe exists in eternity inhabited solely by neutrinos and photons [27, 28] (Fig. 13).

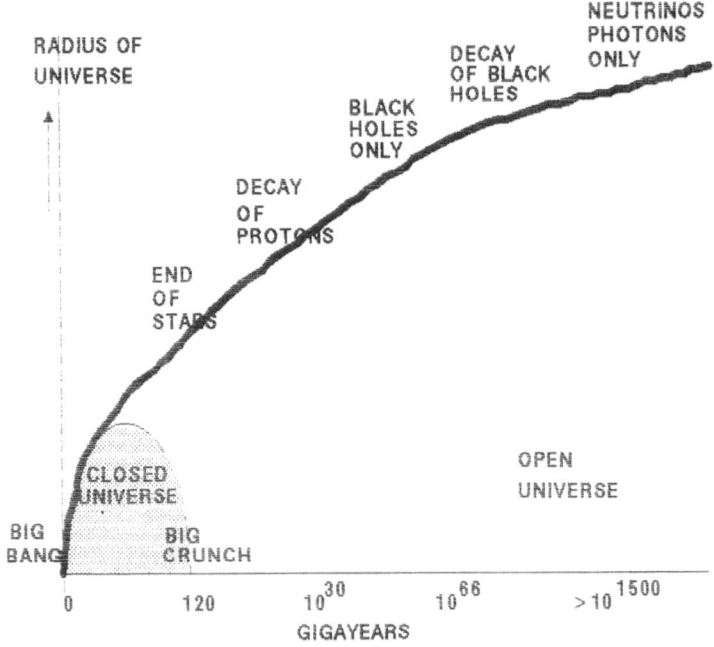

Fig. 13. The very distant future: black holes, neutrinos, photons

Nuclear Boundary Conditions for the Existence of the Realm of Chemical Phenomena

Let us enumerate the conditions which allow the emergence of the realm of chemical phenomena and therefore the existence of living and intelligent beings, not only on this planet but elsewhere and at any time in the Universe. These are the following conditions which must be fulfilled in the realm of nuclear phenomena and in the realms of galaxies and stars:

—the mass of an up-quark must be larger than that of the down-quark, which makes the neutron heavier than the proton and leads in the end to the transformation of a free neutron into proton

—the exact equality of the sum of the electrical charges of 2 up-quarks and 1 down-quark to the charge of the electron

—the instability of Be-8 and the resulting extremely low cosmic abundance of Be, which allows H_2O to exist instead of BeO and at the same time allows the nuclear synthesis of C-12 in relatively large quantities

—the extremely low cosmic abundance of Li, Be and B, for the same reason as before

—the 'coincidence' of the magic nuclear number and the magic electron number for two nuclides: for He-4, which is a noble gas and is chemically inactive, and for O-16, which is chemically extremely active. He and O are the next abundant elements in the Universe, after hydrogen.

—the 'coincidence' of the similar magnitudes of the energy of chemical reaction, of the order of one eV/molecule, and the energy of photons emitted from the long-lived main sequence stars.

Evolution of Terrestrial Matter and Energy

Chemical Reactions in Interstellar Space

Distribution of Cosmic Matter

It may be that the main component of cosmic matter is the neutrino. Assuming that the lightest neutrino has a mass of around 12 eV/c², and taking into account that the number of neutrinos equals around 10^9 per baryon (proton and neutron), the total mass of neutrinos will be around ten times larger than that of baryons. In this case, only 10 percent of cosmic matter exists in the form of baryonic matter, which is able to play a role in chemical interactions.

Only a very small part of cosmic matter, being in the form of atoms, exists in such an environment that chemical interaction is allowed. More than 9/10 of atomic matter is in the form of luminous stellar matter, having a temperature from some thousands to ten million kelvin. Under such condition, no chemical reaction can take place. Part of atomic matter exists in nonluminous stars, such as brown and black dwarfs. Here some chemical reactions seem to be possible. Another part of atomic matter exists in the form of neutron stars, in which components of atomic matter (electrons and atomic nuclei) have been transformed into neutrons and are therefore excluded from chemical interactions. A small, but not insignificant, part of atomic matter 'disappears' into black holes and is therefore inaccessible for optical observation. Only a few percent of cosmic matter exists in the outer regions of luminous and nonluminous stars and is capable of coming into chemical interaction. Of high significance, from our point of view, is the very small amount of this chemically active matter which exists under such conditions where not only gaseous and solid states exists, but also a liquid state. Only an extremely small part of cosmic matter exists at the same time and at the same place in all three states and is able to react chemically.

Gas-Dust Clouds, Chemistry, Energy Sources

Inside the galaxies, in interstellar space, far from the stars, where the temperature is only slightly higher than the temperature of the cosmic background of 2.75 K, exist clouds of cosmic matter in gaseous and solid states—the gas-dust clouds. The average density of cosmic space (including the stellar matter) is around 1 atom per cubic meter. The average density of interstellar clouds is much higher, of an order of magnitude of 10^7 atoms/cub.m. The gas temperature equals 50–100 K; the dust temperature about 10 K.

The gas-dust clouds have a total mass of around 10^{35} kg, that is, 100 000 times the mass of the Sun. There exist very large clouds of up to 10 million solar masses. The elementary composition of the gas-dust clouds reflects the cosmic abundances of the elements, in which the atomic ratio of $(C+N+O)/H$ is about 0.001. A review of the chemical composition of gas-dust clouds is given in Table 13. The stability of these compounds is rather low and the ultraviolet radiation from neighbouring star is destructive (Fig. 14, Table 13).

The gaseous phase consist chiefly of H_2. Some additional abundant species are CO (fractional abundance 10^{-4}) and H_2O (10^{-5}), both abundances with respect to H_2 at 1. Other molecules have fractional abundances significantly lower than these. Ethanol is a trace species in most clouds, with a fractional abundance of 10^{-10}. Nevertheless, for a cloud of 1000 solar masses, the mass of ethanol alone is comparable with the mass of the Earth, while the mass of the C, N and O in the dust is equal to 3 solar masses [29].

Where are gaseous molecules formed? In the clouds themselves, or in external sources? The origin of dust particles could be different. Red giant atmospheres and novae are known to evolve and eject dust particles. Unless there is strong evidence to the contrary, in situ formation mechanisms should be sought. Three major mechanisms of in situ molecule formation from primeval atoms are: 1) gas phase processes, being severely restricted by the low temperatures and pressures to binary collision events; 2) processes on the surface of dust grains, which corresponds to solid catalytic reactions; 3) photochemical processes in the solid grains themselves, due to the ultraviolet radiation which penetrates dense clouds to a sufficient extent and produces radicals in the icy mantles of the grains.

The chemical nature of the dust particles is still not well understood. They seem to consist mostly of silicate with some graphite and metals. Some scientists suspect that the dust particles contain polymerised formaldehyde. The particles are not of a spherical form, with length of around 1 micrometer, which is near to the size of

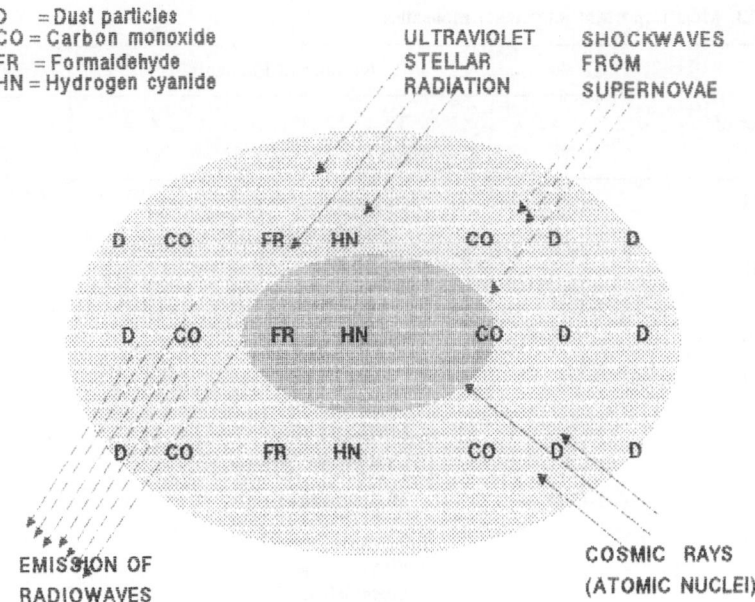

D = Dust particles
CO = Carbon monoxide
FR = Formaldehyde
HN = Hydrogen cyanide

ULTRAVIOLET
STELLAR
RADIATION

SHOCKWAVES
FROM
SUPERNOVAE

D CO   FR HN   CO D   D

D CO   FR HN   CO D   D

D CO   FR HN   CO D   D

EMISSION OF
RADIOWAVES

COSMIC RAYS
(ATOMIC NUCLEI)

Fig. 14. Interstellar gas-dust clouds

terrestrial microorganisms. The dust particles play a double role: they act as catalyst for some chemical reactions and in the shadow of the dust, the destructive action of the ultraviolet radiation is diminished and reaction products could be transported into the interior of the cloud.

Comets and Their Chemistry

The Solar System, and probably other planetary systems (if they could be observed), contain solid bodies with masses of up to 10^{15} kg. These bodies, the comets, are made of frozen H_2O, NH_3, CH_4, CO_2, FeO and other metal oxides. Generally their chemical composition is relatively similar to that of the cold matter in gas-dust clouds [30] (Fig.15).

The comets originated at the same time and probably by similar mechanisms as the other bodies in the Solar System—planets etc. The hypothesis of J. H. Oort claims that a ring exists outside the planets, that is beyond Pluto, which includes more than 10^{12} comets. The total mass of comets is less than the mass of Jupiter. From time to time, due to the gravitational attraction of Saturn or Jupiter, some of the comets are deflected in the direction of the Sun. Another hypothesis has highlighted the gravitational impact of other cosmic objects, such as the invisible, small and cold star called 'Nemesis' which is the partner of a double star system containing the Sun as the other member. Approximately every 26 megayears Nemesis comes into the neighbourhood of the Sun and causes an increase in the number of deflected comets. This is the idea of the 'catastrophe' in terrestrial biogenic evolution [31].

Table 13. Most important interstellar molecules

Number of atoms	Number of Elements			
	2	3	4	5
13		Cyanopenta-acetylene ($HC_{11}N$)		
11		Cyanotetra-acetylene (HC_9N)		
10				
9		Dimethyl ether Ethyl alcohol Ethyl cyanide Cyanotriacetylene		
8	CH_3C_3H	Methylformate		
7		Cyanodiacetylene Vinylcyanide Acetaldehyde Methylamine Methyl acetylene		
6	CH_3CH Ethylene	Methyl mercaptane Methyl cyanide Methyl alcohol	Formamide	
5	Methane	Ketene Methyleneimine Formic acid Cyanamide Cyanoacetylene Butadiynyl radical		
4	Ammonia C_3O C_3H, C_2H_2	Formaldehyde Thioformaldehyde	Hydrocyanic acid Isothiocyanic acid	
3	H_2O SO_2 H_2S CCH	COS HCN HNC HCO HNO		
2	CN HO CC CS CO CH NS NO SO SiS SiO			

When a comet approaches the Sun it has a central solid core, the nucleus, of around 10 km in diameter, which corresponds to a mass of 10^{13} kg, a gaseous and relatively dense coma, plus a rather long but very diffuse tail.

At the beginning of 1986 the comet Halley was observed by a number of space missions. The period of this comet is 76 years and this was its sixtieth rendezvous

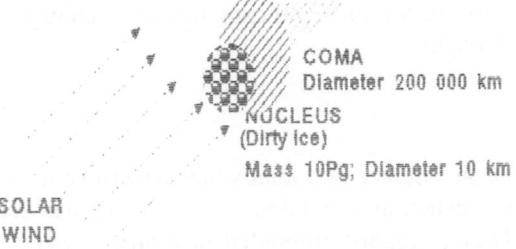

TAIL
Length 100 million km

COMA
Diameter 200 000 km
NUCLEUS
(Dirty Ice)
Mass 10Pg; Diameter 10 km

SOLAR
WIND

Fig. 15. Scheme of a comet: Halley's comet 1986

with Earth. Its core has a length of 14 km, width of 7 km and volume of 500 cubic kilometres and a density of 100 to 300 kg/cub.m. Its mass is thus around 10^{15} kg. The surface is black, with a temperature of around 300–330 K, and is covered by a ceramic-like material, the deeper layers containing 'snow and ice' and numerous solid particles. The albedo is about 4 percent, which means that the nucleus of Halley's comet is the darkest of all known bodies in the Solar System. The coma contains gas and dust particles with a composition similar to that of stony meteorites, with a relatively larger proportion of carbon. In the coma the following entities were identified: H_2O, CH, CN, NH, C_2, HCN, H_2CO, CH_3CN, S_2, NH_3, and N_2, but no Na. Of some surprise was the small concentration of CH_4 and NH_3 and the very small amount of N_2. In the comet's water, deuterium is from 5 to 10 times more abundant than the interstellar average. Also, macromolecules have been observed which correspond to polyoxymethylene (polymerized formaldehyde $(H_2CO)_5$). The dust particles with diameter of 1 μm contain material similar to the chondritic stony meteorites (see next chapter). The elementary composition is: O, C, Si, Fe, Mg and S, in decreasing amounts [32, 33].

Behind the coma could be seen a straight ion tail and a curved diffuse dust tail. The dust tail results from the radiation pressure of the solar radiation on the cometary dust particles. Some dust tails achieve a length of a million kilometers. The ion tail results from the interaction between cometary gas and the solar wind. Its length can reach tens of millions of kilometres with an average density of around a giga ions per cubic metre.

From time to time comets strike the Earth, and are partially vapourised in the upper layers of the atmosphere. The total terrestrial inflow of cometary material is estimated at about 10^9 kg during one megayear. There are hypotheses postulating that, at the present time, the inflow is much more intensive and plays a significant role in the terrestrial balance of water, because the cometary material is mostly ice. Even a conservative assumption claims that 4 w% of the mass of the oceans has a cometary origin. An extreme opinion claims that the total amount of terrestrial water comes from comets. However, on the Moon most impacting cometary water would be lost in ejecta, photodissociation and destruction by the solar wind. Mars would be expected to acquire and retain less impact-delivered water than the Earth, but more than the Moon.

Small comets, even when they reach the surface of the Earth, such as the Tunguska event (1908), do not result in significant chemical changes in the atmosphere, hydrosphere or biosphere.

Meteorites and Cosmic Dust

The Solar System contains a large number of bodies which consist of nonvolatile, nonreactive, solid substances—metals and/or minerals. Their age has been estimated to be one gigayear. They probably originated in a larger, warm cosmic body. Some of them have properties which indicate that they have never been heated above 500 K. A proportion of these solid bodies have a small mass in the range of micrograms to grams, and are mostly vapourized upon entry into the terrestrial atmosphere, at a height of between 70 and 110 kilometres. These bodies have the name 'meteors'. The remaining solid bodies have masses in the range of kilograms up to 10^{12} kg, and fall to the Earth's surface. These are termed meteorites. The chemical composition of meteorites varies from iron-like ('siderites'), through stony irons ('siderolites') to stony meteorites ('aerolites'). The iron meteorites (whose total recorded number is approximately one thousand) consist of metallic iron and nickel, with some sulphides and carbides. The stony meteorites (total recorded number approximately two thousand) consist of silicates of magnesium and iron. Some of the latter type of meteorites, called 'carbonaceous chondrites', contain 25 w% of iron and a significant amount of organic compounds (of around 2 w%), including compounds of high molecular weight and a wide range of aliphatic and aromatic hydrocarbons. In some meteorites, very small crystals of diamond and silicon carbide have been discovered, with sizes of 0.1 to 1 μm. In addition some isotopic peculiarities have been measured [34].

It has been proposed that silicon carbide, such as has been found in the Murray meteorite, originated in the atmosphere of a carbon-rich star. The Sun is not such a star. Does this then mean that in the protosolar nebula another star of the carbon-type had a significant influence on our Solar System? Or are these the trace remains of a supernova?

The most extensively studied chondritic meteorite is the 'Murchison', which fell in the State of Victoria in Australia near the town of Murchison, on September 28, 1969. Within only a few days of its landing, the first samples were examined in laboratories using advanced techniques. The meteorite contained 2 percent by weight of carbon and 0.16 percent by weight of nitrogen. Most impressive, however, is the amino acid content (Table 14).

All these amino acids are racemic and the products of abiogenic spontaneous chemosynthesis in the gas-dust cloud before the emergence of the protosun.

The terrestrial annual inflow of meteoric matter at the present time equals roughly 20 Gg (20 000 ton). The Earth collects far more mass in the form of cosmic dust (particles with masses of the order of magnitude of milligrams) than it does in the form of the larger particles, the meteorites. In the past, especially 3.5 Gyr ago, the inflow was very large and significantly influenced the mass and the chemical composition of the evolving Earth. The total sum of the meteoritic material accumulated during the past 3.5 Gyr, assuming an inflow at the present level, corresponds to a global layer with a height of around 0.1 m.

Micrometeorites (or cosmic dust) of less than 1 mm in size have been collected in deep-sea sediments, on the Greenland ice cap, and in the lower stratosphere. It is probable that cosmic dust is the product of collisions between asteroids, but cometary origin cannot be excluded.

Table 14. Organic substances in the Murchison meteorite

Species	Abundance
Acid insoluble carbonaceous phase	1.3–1.8%
Carbonate and CO_2	0.1–0.5%
Hydrocarbons	
Aliphatic	12–35 ppM
Aromatic	15–28 ppM
Acids	
Monocarboxylic (C2-C8)	170 ppM
Dicarboxylic (C2-C9)	+not measured
Hydroxy (C2-C5)	6 ppM
Amino acids	10–20 ppM
Glycine	6 ppM
Glutamic acid	3 ppM
Alanine	3 ppM
Valine	2 ppM
Proline	1 ppM
Aspartic acid	1 ppM
Alcohols (C1-C4)	6 ppM
Aldehydes (C2-C4)	6 ppM
Ketones (C3-C5)	10 ppM
Ureas	20 ppM
Amines (C1-C4)	2 ppM
N-heterocycles	
Pyridines and quinolines	0.01 ppM
Purines	1 ppM
Pyrimidines	0.05 ppM
Poly-pyrroles	<1 ppM
Sum	1.43–2.35%
Total carbon	2.0–2.5%

Protoplanets and the Evolution of Planets

Origin of Solar Planets

The present state of hypotheses concerning the emergence of the planets does not give convincing answers to the following questions:

—are planets formed during a catastrophic event, such as the explosion of a supernova in the neighbourhood of a young double-star system? There exist some traces in our Solar System of the influence of a supernova, for example in the existence and the isotopic composition of the heavy radioactive elements Th and U.

—are planets formed as a 'normal' product of the evolution of the protosolar gas-dust cloud? The large number of planets and their satellites, and the huge number of asteroids, meteorites and comets are evidence of such a mechanism of emergence of a planetary system.

—do planets originate from a certain type of star, and only during a certain period of their evolution?

—how numerous are these stars and what is the probability that a planetary system will be formed?

However, up until today we have no direct evidence of the existence of another planetary system than our own, even if some candidates for protoplanetary discs around some neighbouring stars do exist.

Origin of Protoplanetary Clouds

Too few facts are known (we have access to only one planetary system, which is relatively old—4.5 Gyr) and the hypotheses are too weak to be able to formulate a coherent general theory of the emergence and evolution of planetary systems. A simple description of these processes could be as follows [35, 36].

In the protosolar gas-dust cloud, the accretion of small, cold dust particles, with a diameter of around 1 μm, results in the formation of large solid bodies, due to gravitational attraction. The average chemical composition of the whole gas-dust cloud is similar to the chemical composition of the protosun, which is shown in Tables 1–12. The chemical composition of the dust grains is the crucial parameter. The emerging larger bodies have a diameter of order of magnitude of 1 m up to 1 km, and are called 'planetesimals'. The integrity of these bodies is controlled by forces other than gravitation, for example solid state forces. The accretion proceeds very slowly, over around 100 Myr, and acts selectively. The less volatile components, such as He, Ne, Ar, and H, and even N, remain in space and later dissipate outside the Solar System. The less volatile, such as H_2O, CH_4, NH_3, H_2S, CO_2 and of course graphite, all metals and metallic oxides, sulphides, etc., stay in the planetesimals. This is the first large chemical differentiation. For example, helium is 10^{14} times and krypton 10^7 times less abundant on Earth than within average cosmic matter. The collapse of a swarm of 1 km-planetesimals is thought to be fairly short, of the order of 100 years or so.

The dynamics of accretion of planetesimals (diameter 1 km) into planetoids (tens or hundreds of kilometres in diameter) are complex. The accretionary growth of planetoids probably begin as a rather gentle aggregation of many planetesimals. The further growth of most of these planetoids into planets would be strongly influenced by their orbital characteristics. If accretion occurred rapidly, the decay of surviving short-lived Al-26 (half-life of 730 000 years), being the product of a supernova explosion, may be capable of melting planetesimals as small as a few kilometres in diameter. Slower accretion (a few million years and a few kilometre in diameter) would limit the radioactive heat sources of 1000–1300 K would destroy early-formed organic compounds, but it is possible that the internal heat generation would have indirect effects on the chemical processes of the near-surface layers.

After a million years of continued accretion of planetoids, a planet emerges. The heat generated due to collisions and to the radioactive decay of long-lived nuclides, results in an increase in the temperature of the deeper planetary layers. The heat flux of the radioactive nuclides would be roughly 4 times higher at present. The further accumulation of planetesimals forms the large body of the Protoearth with a mass of about $5 \cdot 10^{24}$ kg. The temperature in the centre of the Protoearth would probably be 1750 K. Metallic iron, together with iron oxides and sulphides, being denser move towards the Earth's centre. Molten silicates including radioactive nuclides K-40, Th-232, U-235 and U-238, being lighter, move towards the outer terrestrial layers. This second, large chemical differentiation is called the 'iron catastrophe'.

During the further evolution of our planet, chemical differentiation proceeds in parallel with homogenisation.

Planets, Chemical Composition

In spite of the fact that the evolution of all planets is not well enough understood there are some obvious parameters which decide the evolutionary path of a planet: its distance from the Sun, that is, the influx of solar radiation and the resulting temperature of the planetary surface, the initial mass of the protoplanet, its initial chemical composition, etc. Different combinations of parameters result in the emergence of different planets (Table 15).

The Planet Earth

Terrestrial Abundance of Elements

To repeat, the chemical composition of the Universe is as follows. In each 1000 atoms, the number of hydrogen atoms equals 920, helium atoms 78, oxygen atoms 0.6, carbon 0.3, nitrogen 0.08, iron 0.03, and all of the other more than 80 elements less than 0.5. Because the two most abundant elements, hydrogen and helium, originated from the first period of the evolution of the Universe, it can be claimed that the Universe is still relatively young, in spite of its 15 Gyr age.

The chemical composition of the Sun shows clearly more medium and heavy elements, including Th and U, which is evidence that the Sun is a star of the third generation.

The chemical composition of the Earth as a whole shows another structure. Because the solid sphere of the Earth contains 99.977 percent of the total terrestrial matter, we can consider this part of the total Earth to be quite a good representation of the whole. The main component of the Earth, expressed by weight, is iron. This metal is the most abundant metal in the Universe, in the Sun and in our planet. It is, at the same time, the most stable nuclide, that is, the nuclide with the largest binding energy—8.8 Mev per nucleon (see Fig. 4 and Table 5). Iron is the very 'ash' of all nuclear processes, of the fusion of the light elements and of the fission of the heavy elements. This is the real reason for the relatively large cosmic, solar and terrestrial abundance of iron. It is not a trivial fact that, after the epochs of stone and bronze tools, we are still living in the iron era. The most abundant structural material of our civilisation is the most stable, and cosmically the most abundant nuclide [38].

Table 15. Chemical composition of the planets. cub.m = planet mass relative to the mass of the Earth; d = mean density in 1000 kg/cub.m; t = average surface temperature in °C or K; p = atmospheric pressure in bars

Planet	Spheres			
	Solid	Liquid	Gaseous	Biotic
Mercury m = 0.054 d = 5.3 t = 550 °C	Silicates of Al, Fe, Ca	No	No	No
Venus m = 0.814 d = 4.95 t = 470 °C	Silicates of Al, Fe, Mg, Ca with metallic core Volcanic activity	No	Very dense p = 100 90% CO_2 1% N_2, SO_2, HCl 0.4% H_2, H_2SO_4	No
Earth m = 1.00 d = 5.52 t = −15 °C	Silicates of Al, Fe, Mg, Ca with metallic core: Fe, Ni	Water	p = 1 778% N_2 21% O_2 0.03% CO_2	Yes
Mars m = 0.108 d = 3.95 t = 260 K	Silicates of Al, Fe, Mg	No	p = 0.01 99% CO_2 1% N_2 CO_2	No
Asteroids (Total) m = 0.001 d 1 ≪ 5 t very low	Solid Silicates	No	No	No
Jupiter m = 317.45 d = 1.33 t = 127 K	In centre stony core with m = 14 and solid hydrogen?	Probably liquid ammonia and methane	Very dense 25% H_2, 15% He NH_3, CH_4, H_2O, SiH_4, H_2S, C_2H_6 CH_3NH_2, C_2H_4, HCN	No

Table 15. (contd.)

Planet	Spheres			
	Solid	Liquid	Gaseous	Biotic
Saturn				
m = 95.06	In centre	No	Dense	No
d = 0.68	stony core		88% H_2,	
t = 97 K	with m = 16		11% He, 0.1% H_2O	
			0.1% CH_4, NH_3	
Uranus				
m = 14.5	In centre	No	p = 200	No
d = 1.56	stony core		H_2, CH_4, He	
t = 58 K	and ice			
Neptune				
m = 17	In centre	No	Yes	No
d = 2.27	stony core		H_2, CH_4, He	
t = 53 K				
Pluto				
m = 0.002	Solid 'gases'	No	Very low	No
d = 2	NH_3, CH_4		density	
t = < 40 K				

The next most abundant component of the Earth as a whole is oxygen, with approximate abundance of 30 w%. Oxygen is, after hydrogen and helium, the third most abundant element in the Universe and in the Sun. Oxygen-16 is a very stable nuclide, containing the magic number of protons, 8, and the magic number of neutrons, also 8. (see Table 9). It is also important to see that oxygen-16 originated from helium burning, according to the relationship:

$$3(\text{He-4}) \rightarrow \text{C-12}; \quad \text{C-12} + \text{He-4} \rightarrow \text{O-16}.$$

The next most abundant element on the Earth as a whole is silicon, with about 15 w%. The origin of silicon can be simply represented in the following way:

$$\text{O-16} + 3\,(\text{He-4}) \rightarrow \text{Si-28}.$$

The fourth element in terms of abundance is magnesium, with around 13 w%; also a product of helium burning. Together these four elements account for more than 90 percent by weight of the Earth.

The Earth's crust has a different chemical composition from that of the whole Earth, with oxygen as the most abundant element, at 46% by weight. This corresponds to 63 atoms of oxygen in each 100 atoms of the crust. Because of the anionic state of oxygen in the silicates, the volume of oxygen in the crust reaches around 93 percent by volume. The remaining 7 percent by volume accounts for all other elements. In reality, the Earth's crust should be called the 'oxygen sphere'.

From all of this, it must be clear that the Earth has lost an enormous amount of the volatile elements H, He, Ne, and also C and N. The amount of C and N, even if much smaller than can be expected from their cosmic abundance, is still large enough to significantly influence chemical interaction on the terrestrial scene and form the basis of terrestrial life.

In the chemical composition of the Earth, the heavy elements seem to play a rather insignificant role. The cosmic abundance of thorium and uranium equals 1.3 ppT (parts per Tera) and 0.3 ppT, respectively. In the Earth, the abundance of both heavy elements is much higher, and is around 12 ppM and 4 ppM respectively. The role of heavy nuclides in the energy balance of the Earth is very important in spite of their small abundance. The heat flux coming from the deeper layers of the Earth, the energy of volcanic activity, the processes of degassing of the internal layers of the Earth, the drift of the continents during the life of the Earth, the energy of earthquakes, etc, all result to a large degree from the spontaneous decay of the heavy elements, thorium and uranium. These elements are the products of nuclear processes in the outer layers of a supernova explosion which influenced the protosolar gas-dust cloud 5 Gyr ago.

The heavy elements are very important carriers of free energy in man-made energy sources. However, one should not underestimate that the fission products of the heavy elements are the source of extremely dangerous radioactivity.

The Stability of Terrestrial Matter

The Earth contains all the stable nuclides which exist in the Universe, without exception. This statement is not as trivial as it seems to be. It is possible that, somewhere in the Universe, exist planetary systems surrounding stars of the second generation which do not contain heavy elements such as Th and U. It is also not so obvious that all stable nuclides are contained in the outer layers of a planet, that is, in the crust, in spite of rather extensive and intensive chemical differentiation during its gigayears of evolution.

It must be stressed that most technological processes do not influence the stability of nuclides, but only change the molecular environment of the elements. At the present time, the two exceptions to this are: the irreversible fission of the heavy elements—the naturally occurring U-235 and the man-made Pu-239 and Pu-241, and, in the future maybe U-233 in nuclear power reactors,—and secondly the breeding process, which transforms naturally occurring Th-232 and U-238 into U-233 and Pu-239 and Pu-241, respectively. In the future the fusion process will also play a significant role. In particular, the irreversible transformation of naturally occurring deuterium and man-made tritium into helium will be of significance. The breeding of tritium from lithium will also be important. All other processes, mostly chemical, thermal or mechanical, do not influence the stability of the elements (Fig. 16).

Terrestrial material is held together by gravity. Only small amounts of matter from cosmic space reaches the Earth in the form of larger cosmic bodies, such as meteorites and comets. The gravitational field is large enough to prevent the escape from the terrestrial atmosphere of almost all molecules, with the exception of the lightest, such as H_2 and He. However, even this fact has an impact on the chemical environment. The small amount of water which is photolysed by solar ultra-violet

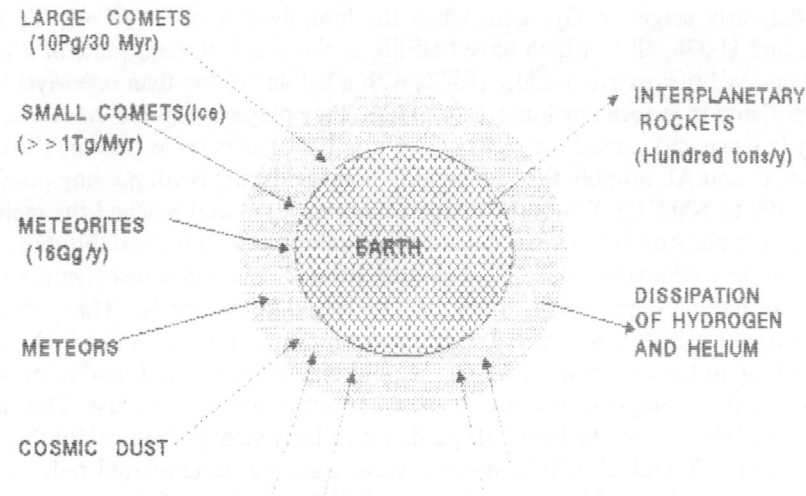

Fig. 16. Terrestrial matter in the gravitational field

radiation produces H and O. Small amounts of H escape, but the oxygen remains in the terrestrial atmosphere and increases the amount of unbonded molecular oxygen.

Other than these random escapes, only artificial objects—space rockets, sent aloft by man—can leave the gravitational field of the Earth. The high price which must be paid in energy terms to separate those rockets from the Earth is well-known, and it seems fairly certain that, in the future, the cost of the transport of materials from other planets, comets or asteroids to the Earth will also be economically and energetically prohibitive. This is especially self-evident when it is remembered that all the stable and quasi-stable elements already exist on this planet in rather large amounts and are rather readily accessible.

Chemical Changes in the Past

The Earth is 4.5 Gyr old, but the oldest known continental crust has been dated at 3.7 and 3.8 Gyr. Events during the first 0.7 to 0.8 Gyr remain obscure. There are some questions which are therefore raised. Did the continents commence their formation 3.8 Gyr ago? If so, what was the thermal state before, and why did the continents wait so long to appear when the atmosphere and part of the hydrosphere seem to have formed much earlier? How has the continental mass changed during the life of the Earth? Has it remained constant or has it grown or decreased continuously, for example?

The history of the very early Earth must have been dominated by the arrival of extraterrestrial bodies on the surface. The result of these arrivals was among other things an increase in temperature, which caused the intensive degassing of the Earth's interior, taking the volatile components away. This is one of the numerous mechanisms of chemical differentiation of planetary material.

In the early stages, 4 Gyr ago, when the long-lived nuclides (K-40, Th-232, U-235 and U-238; all of which have half-life of the order of magnitude of a Gyr) and short-lived radioactive nuclides (Al-26, with a half-life of less than one Myr) were 'younger' and therefore much more abundant, they played a significant role in the energy balance and caused the melting of a number of silicates, especially those of Ca, Na, K and Al, notably the low melting point feldspars (with melting point as low as 700 to 1000 °C). When the temperature increased and reached the melting point of metallic iron (1535 °C), a catastrophic sinking of iron toward the core took place and the radioactive 'fuel' began to concentrate in the outermost layers, where the heat that was generated could be conducted more easily. The chemical differentiation which took place during the evolution of the Earth resulted in the iron sinking to the core and Si, Al, Ca, K, and Na, in the form of oxides, moving upward, with consequent increase in their concentration in the crust. The same paths were followed by the heavy elements which have strong chemical interaction with oxygen: Th and U, which at the present time are accumulated only in the crust. The heavy but noble metals, Au and Pt (also products of nuclear synthesis during a supernova explosion) probably mostly sank into the core. This is a good example of how the two elementary forces, gravitation, which depends upon the mass of the planet, and the electromagnetic force, in the form of chemical interaction, which partially depends on the balance of energy and the resulting temperature, influenced the history of the solid sphere of the planet.

Terrestrial Flow of Energy and Matter

Energy Flow: in the Past, Today and in the Future

The past, present and future thermal regimes of the lithosphere depend critically on estimates of the bulk composition of the Earth, especially the total abundance of the long-lived, heat producing radioactive elements Th and U and radioactive K-40. The U content is no less than 15 ppG and certainly no more than 30 ppG. The ratio of Th/U is well-known and equals 3.8. The ratio of K/U equals 10^4. Potassium is the eighth most abundant element in the Earth's crust, with a concentration of 1.68 w%. Natural potassium consists of two stable isotopes, K-39 and K-42 and one radioactive isotope, K-40, in a proportion of about 0.01%, which corresponds roughly to 2 ppM. Its half-life is 1.3 Gyr [38].

If the estimate of 20 ppG for U and the appropriate amounts of Th and K-40 are correct, the total heat flux out of the Earth exceeds present-day heat production from radioactive nuclides by a factor of two. One-fourth to one-third of the Earth's uranium is present in the continental crust, the rest is distributed in the mantle. The concentration of U may be only 5 ppG in the upper mantle. The concentration of Th in the crust is 10-13 ppM. The radiogenic heat flow is estimated to be in the range of 25 ± 4 TW.

The remainder of the heat flow also estimated as being of the order of 25 ± 5 TW, results from the secular cooling of Earth, that is, from the primordial heat sources. Such a large distribution due to secular cooling is most easily achieved if mantle

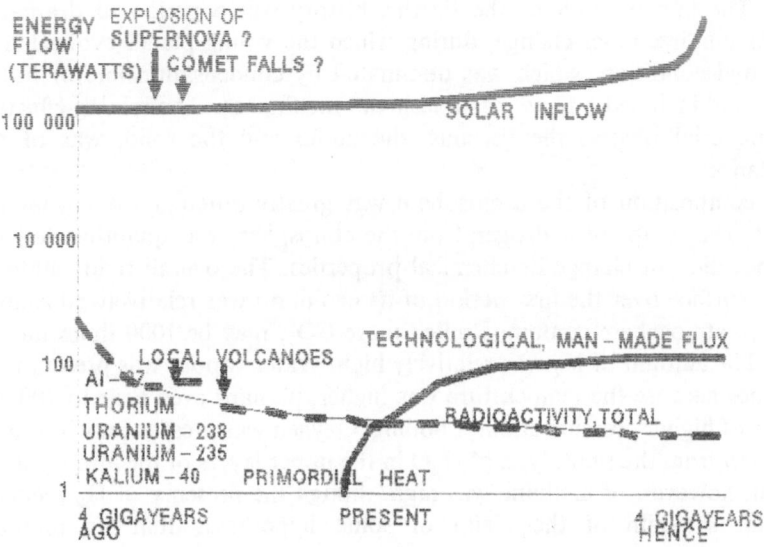

Fig. 17. Energy flow: in the past, today and in the future

convection is layered. Even then the temperature of the upper mantle must have changed by about 200 °C (from 1550 to 1350 °C) during the last 3 Gyr (Fig. 17).

Abiogenic Chemical Evolution of Terrestrial Matter

The Conditions of Early Chemical Evolution

Roughly 4 Gyr ago, the primitive Earth was the scene of very intensive and extensive chemical processes—the spontaneous abiogenic synthesis of very numerous and rather complex molecules. The conditions for chemical abiogenic synthesis are far from the conditions ruling on the Earth at the present time. Here, in brief, are the most important agents acting 4 Gyr ago.

The Sun was at the same evolutionary phase as today, that is it was a main sequence star. However, at this time (4 Gyr ago) the Sun was only at the beginning of this phase, while today it is half-way through it. The surface temperature of the Sun was markedly lower than it is today. The solar radiation flux toward the Earth was about 20 percent lower than it is today; that is, approximately 1.1 kW/m². Also the spectrum of solar radiation was shifted to the red end, in comparison with the present time. The flux of ultra-violet radiation was consequently lower, which means that its impact on chemical processes was reduced. It must be taken into account, however, that the primitive atmosphere did not contain molecular oxygen and therefore ozone was also absent. This situation allowed solar ultra-violet radiation to reach the surface of the Earth [39, 49].

Volcanic activity was probably very intensive and extensive. The falls of large and medium-size cosmic bodies (comets, asteroids, and meteorites) had a significant impact, both on the energy balance and on the material balance of the early

Earth. The first 0.7 Gyr of the Earth's history was a period of dramatic and perhaps unidirectional change, during which the young planet evolved from its primordial condition, which was dominated by condensation and/or accretion. This period included a phase of widespread melting, and the mixing effects of all three material phases, the gaseous, the liquid and the solid, was of utmost importance.

The composition of the atmosphere was greatly different from what it is at present. The escape of hydrogen from the atmosphere was quantitatively significant, but did not change its chemical properties. The overall redox state of the Earth's surface over the first period of its evolution was relatively constant. The primitive atmosphere contained much more CO_2, may be 1000 times more than today. The amount of N_2 was relatively high. Water vapour was present in lower quantities because the temperature was higher, perhaps even near to 100 °C, but because of higher pressure still not boiling. Oxygen was present only in traces and originated from the photolysis of H_2O in the upper layers of the atmosphere. The amount, however, of methane was high, though the presence of H_2 seems to be doubtful, in spite of the claim of some hypotheses that its presence is obvious.

The mass of the present atmosphere is equal to $5.1 \cdot 10^{18}$ kg, which corresponds to 0.89 ppM of the total mass of Earth. The primordial atmosphere probably had a mass 100 times larger, mostly due to the CO_2.

The mass of the early hydrosphere was probably smaller than at present. The amount of surface water seems to have increased with time because of the degassing due to volcanic activity, and perhaps due to the arrival of icy comets [41]. However, it seems that the continental areas were significantly smaller. Both these statements can be accommodated by the assumption that the Archean ocean was shallower than the present oceans. It seems that the influence of the moving continents on the emerging oceans (of the Atlantic type) has been underestimated [37, 43]. It must be noted in passing that the lithosphere in a complete by lifeless desert undergoes different geochemical processes than at present.

Primordial Soup; Abiogenic Synthesis

The primordial atmosphere probably included many molecules, such as hydrides of C, N, P, and S, and oxides of C, N, P, S, etc. (see Schidlovski Handbook, Vol. 1, Part A, Page 1). The primordial hydrosphere contained a rather concentrated (at least locally) aqueous solution of hydrides and oxides of the 'life-carrying' elements, called the 'primordial soup'. The abiogenic, spontaneous reaction between these two compounds, with the influx of free energy, resulted in the synthesis of relatively large numbers of amino acids, sugars, nucleotides, phosphoric acids, and iron salts. The free energy was in the form of solar ultra-violet radiation, radioactive energy, local volcanic heat, and the shockwaves from falling extraterrestrial bodies.

It should be taken into account that the hydrosphere in the initial period of terrestrial evolution contained at least one order of magnitude less water than at present; that is, probably 10^{20} kg. the temperature was much higher, reaching

100 °C, but because of higher atmospheric pressure, mostly due to CO_2, far from its boiling point.

Under these conditions, some polymerization processes are also possible. Spontaneous polymerization of charged amino acids results in the synthesis of simple peptides. Also, spontaneous randomly structured synthesis of simple, short ribonuclei acids is not unlikely. The latter class of compounds not only play the role of a matrix for the polymerization of amino acids, but also have some catalytic properties. Some clays could probably also play the role of being matrices for the spontaneous polymerization of amino acids. These processes could have taken place during the drying-out of small water reservoirs in the neighbourhood of hot lava.

Another way of producing polymerized macromolecules could be the hydrolysis of spontaneously polymerized simple molecules, such as HCN, into an HCN-oligomere, resulting in the synthesis of amino acids (e.g. Gly, Ala), purine (e.g. adenine), imidazole and pyrimidine (e.g. 4-5-dihydroxypyrimidine) [44].

The material basis for further steps of chemical evolution was thus prepared.

Chemical Impact of Comets and Meteorites on the Terrestrial Environment

Not only terrestrial conditions are of significance in the preparation of the 'primordial soup'. The impact of extraterrestrial phenomena is of greatest importance. The probability of the influence a supernova explosion, close enough to the Earth (e.g. some light years away) to be a potential source of environmental effects on the Earth's surface, seems to be very small, but not negligible.

The largest cosmic bodies which from time to time collide with Earth are the comets and meteorites. The mass of the larger examples is estimated on the low side and assuming a rocky-iron asteroid with a radius of 3 kilometers, to have a mass of $500 \cdot 10^{12}$ kg and a velocity of 20 km/s, which is typical of an object orbiting in the inner Solar System. On the high side, is an ice-rich comet coming in from the far reaches of the solar system with a velocity of 65 km/s, and having a radius of 14 km and mass of $13 \cdot 10^{15}$ kg. The chance of collision with Earth cannot be considered to be negligible. Particularly in the early stages of the Solar System, collision could have had an impact on the chemical evolution of the atmosphere and hydrosphere, and maybe on the origin of life. It should be taken into consideration that the mass of the larger bodies is of order of magnitude of $10^{14 \pm 1}$ kg and the mass of the present terrestrial atmosphere is $5 \cdot 10^{18}$ kg. The speed of entry of meteorites ranges from 13 to 21 kilometres per second. The kinetic energy of such falling bodies with a mass of 10^{13} kg reaches 10^{22} joules, which corresponds to 2 days of the total terrestrially absorbed input of solar radiation. A 10 km diameter meteorite may produce some Pg of NO, enough to destroy almost all the ozone layer.

The present collision rate between the Earth and such a large cosmic body as an asteroid of around 1 km diameter is estimated to be 6 per Myr. Asteroids with 10 km diameter collide with the Earth once per 50 Myr. The collision rate of comets is lower than that of asteroids. Brief, but intense, comet storms have probably occurred once every 100 Myr.

The most recent comet which is believed to have collided with Earth was 'Tunguska'. On June 30, 1908, near the Siberian village of Tunguska, a rather small comet with a mass of $3 \cdot 10^7$ kg struck the Earth. The energy released during this impact has been estimated at 50 PJ. Many people observed this cosmic event and many scientific measurements of it were made. Even this event, however, had only a rather local and limited effect. The probability of the impact of such a small comet on the Earth is about one every thousand years. A collision of a large comet (100 to 1000 times heavier) with the Earth results not only in the release of enormous amount of energy, but also in the effect of the extraordinary amounts of rather volatile and extremely active substances, including ions and radicals, which are generated. All these extraterrestrial phenomena have played a decisive role in the emergence of the 'primordial soup'.

Evolution of the Terrestrial Hydrosphere in the Past, Today and in the Future

The present hydrosphere weighs $1.4 \cdot 10^{21}$ kg. Part of this exists in solid form in the cryosphere, with a mass of $4.6 \cdot 10^{19}$. This total amount of water corresponds to approximately 230 ppM in relation to the total mass of Earth, and the amount of ice corresponds to 8.2 ppM.

The origin of seawater could be degassing and/or accretion of 'icy' comets. In the past the world's ocean had a smaller mass and covered the whole surface. The continents also had a smaller surface than at present. The global sea level changes during periods of 10 to 20 Myr between 150–200 m (Fig. 18).

Evolution of the Terrestrial Lithosphere in the Past, Today and in the Future

The present lithosphere (here as total solid-state matter on the Earth, including all the deeper layers, even if they are in liquid form) has a mass of $5.5 \cdot 10^{24}$ kg; that is, about 99.977% of terrestrial matter. The crust corresponds to $2.4 \cdot 10^{23}$ kg.

Volcanic activity (lava and ashes) produces some $3 \cdot 10^{12}$ kg/year. The fact that volcanic activity and the emergence of new rock material on the Earth's surface are directly connected with relatively large emissions of very active chemicals, seems to be underestimated. One cubic kilometre of basalt ejects approximately $2 \cdot 10^9$ kg of sulphuric acid and $5 \cdot 10^7$ kg of hydrochloric acid into the atmosphere and/or hydrosphere (Fig. 18).

Water as Erosion Factor in the Lithosphere

Water plays a significant role in the transport of the components of the solid sphere to the Earth's surface. The total global amount of solid material transported by rivers has been estimated at approximately $9 \cdot 10^{12}$ kg. This amount corresponds to the average annual denudation during the Tertiary Era. The annual amount of total water flux in the world's rivers is some $39 \cdot 10^{12}$ cub.m. The average amount of solid particles transported in the rivers is around 0.25 kg per cubic metre of water.

The total amount of sediments on the Earth is estimated at $3.2 \cdot 10^{21}$ kg, which corresponds to a world-wide thickness of about 2.2 km. In the past the average rate of erosion due to water and wind have been around 0.01 mm/year. The present

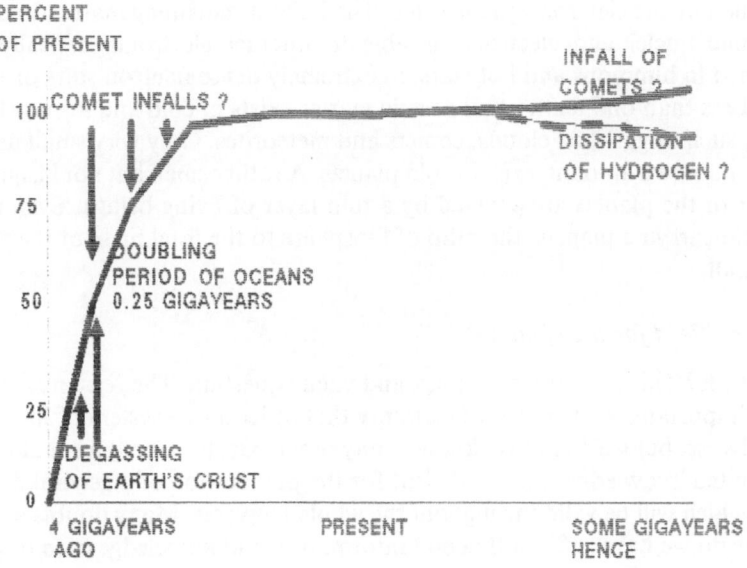

Fig. 18. Evolution of the hydrosphere

erosion rate seems to correspond to 0.03 mm/year, reaching 0.06 mm/year in North America.

The present total rate of erosion and of sedimentation equals 20 Pg/year. Approximately 3/4 of recent sediments are derived from the weathering of old sedimentary rocks, and 1/4 from the weathering of igneous and metamorphic rocks (Fig. 19, Table 16).

Terrestrial and Extraterrestrial Life

General Definition of Life

Life; Coupling of Matter and Energy

Whatever the definition of life is, one thing seems to be sure, life is an extremely complex phenomenon and therefore all four elementary forces and all elementary stable particles must be involved.

All particles (bricks), stable and unstable, and all field particles (mortars) are influenced by the Gravitational force. All particles are influenced by the Weak Force. Heavy particles, baryons, are ruled by the Strong force. But far from all the particles are influenced by the Electromagnetic force, only those which are, or can be, electrically and/or magnetically charged. As has been previously mentioned, probably more than 90 percent of cosmic matter exists in a form which is not influenced by the electromagnetic force. Most matter probably exists in the form of neutrinos or other exotic particles, which are electrically neutral and are not

controlled by the electromagnetic force. The bulk of remaining matter, in the form
of atomic nuclei and electrons, is able to interact electromagnetically but is
contained in luminous and hot stars, in extremely dense neutron stars or in black
holes. Less than one-tenth of all cosmic matter exists in cold and averagely dense
bodies, such as gas-dust clouds, comets and meteorites. Only very small amounts,
still an unknown amount, exist in cold planets. A rather small but not insignificant,
number of the planets are covered by a thin layer of living beings. Even in these
latter life-carrying planets, the ratio of biosphere to the total mass of the planet is
very small.

Life: The Need for a Definition

What is life? This is a rather complex and vague question. The reason is as simple
as it is important; we know with certainty that at least one system of life exists in
the Universe, but we do not yet know of any other extraterrestrial form of life. This
very limited knowledge is not sufficient for the preparation of a general definition
of life, which will be valid throughout the whole Universe. Many doubts arise. For
example do we have sufficiently good information and knowledge even to prepare
a general theory for terrestrial life? Nevertheless we will attempt to do this [45, 46].
There exists limitless possibilities for the formulation of opinions, hypotheses or
even faiths, concerning the following assumptions:
—the abundance of life;

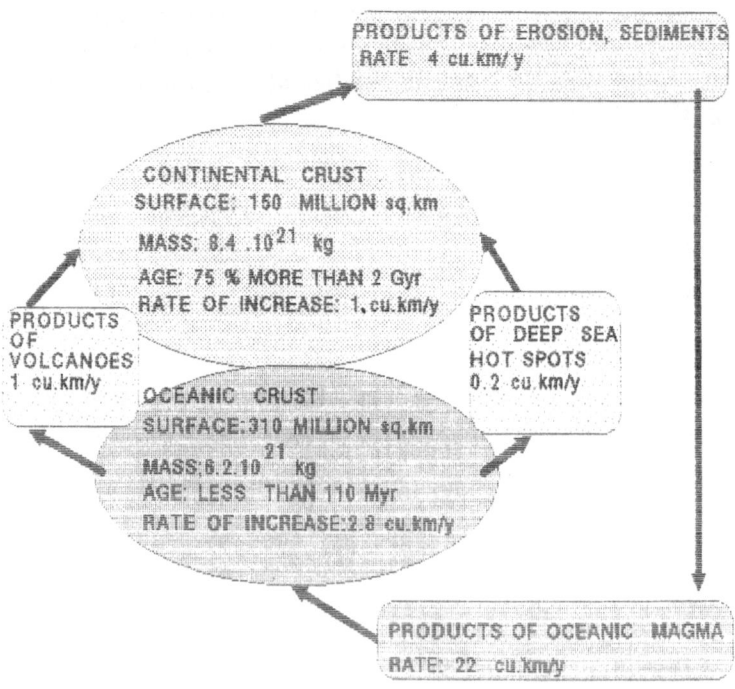

Fig. 19. Matter recycling in the lithosphere

Table 16. Chemical composition of the earth's crust at present. Contents in grams per 1 Mg (1 ton) of crustal rocks, volume: 0.357 cub. metre [38].

1 000 000		10	10 Boron
	446 000 Oxygen		
	227 000 Silicon		
			3.5 Uranium
			1.5 Tin
100 000		1	
	81 000 Aluminium		
	58 000 Iron		All
	36 000 Calcium		other
	28 300 Sodium		elements
	27 700 Magnesium		with
	16 800 Potassium		atomic
10 000		0.1	number
	8 600 Titanium		41 < Z < 87
			0.020 Mercury
	1 400 Hydrogen		
	1 100 Phosphorus		
1000		0.01	
	950 Manganese		
	465 Fluorine		0.005 Platinum
	425 Barium		
	375 Strontium		
	300 Sulphur		0.002 Gold
	200 Carbon 190 Chlorine		
	165 Zr, 135 V		
100	100 Cr, 90 Rb	0.001	
	80 Zn, 72 Ni		
	60 Ce, 55 Cu		
	35 Y, 30 La		
	28 Nd, 28 Co		
	22 Sc, 20 Nitrogen		
	20 Nb, 15 Ga		
	12 Thorium		
10	10 Lead	0.0001	

a) terrestrial life is unique in the Universe

b) life is present in other parts of the Universe, or in other parts of the Galaxy, or even other parts of the Solar System

—the origin of life;

a) life originated spontaneously on this planet without the interaction o any other extraterrestrial living system

b) terrestrial life is the product of the accidental activity of extraterrestrial life.

c) terrestrial life is the product of the deliberate action of extraterrestrial intelligent beings

d) life is the product of currently unknown natural or even supernatural forces

These ideas have been disputed for a very long time and it is not expected that any solution will find general acceptance in the near future. It is unlikely that extraterrestrial living organisms could have reached the Earth, either as spores driven by radiation pressure from another star or as living organisms embedded in a meteorite. As an alternative to these 'nineteenth-century mechanisms', some prominent scientists (including a Nobel prize-winner in the field of DNA research) during the past 15 years have considered 'Directed Panspermia', the theory that organisms were deliberately transmitted to the Earth by intelligent beings from another planet. They conclude that it possible that life reached the Earth in this way, but the scientific evidence for this hypothesis is inadequate. Other scientists have proposed a model in which bacteria are reproduced as the byproduct of star formation in stellar environments where reproduction can occur. Due to the subsequent star formation, a fraction of the bacteria produced are expelled into interstellar space. Some bacteria die, but some survive to become incorporated in further star formation processes. The Earth is perpetually embedded in a halo of cometary material. Small particles of the size of bacteria and viruses could land 'softly' without their viability being destroyed by flash heating.

Nevertheless, the case presented in this chapter relies on the premise that life is not a local and solely terrestrial, but a cosmic, phenomenon and originated as a result of the spontaneous evolution of chemical processes. All this allows us to try to define life on the basis of the present state of knowledge of the natural sciences, knowing that this is valid only at this present moment.

It must be highlighted that the definition of life presented here is not the generally accepted one, but is only a personal point of view which can help to solve our problem. To do this, we will use the scientific language of physics. We must use such a description of the phenomenon of life, so that all specifically local peculiar properties of terrestrial life can be excluded [47].

Spontaneity and Universality of Life

A general definition of life must begin with the most general statements. Life is a universal, that is not a local and not a terrestrial or Solar System specific, phenomenon and not specific to our Galaxy (the Milky Way), but cosmic and universal. It is not limited in time but nevertheless must emerge at some point in time later than the beginning of the Universe itself. Life should emerge at numerous points within the Universe, fully independently, and evolving also independently, at least during the initial period of evolution. However, this does not mean that in all cases life must be of the same form, and it is not even clear if it must be similar or analogous.

The above-formulated general property of life could also be formulated in such a form that life can be said to emerge spontaneously and not be coupled to local peculiarities. This does not mean that at every point in the Universe, and at every point of time, the probability for emergence is equal. We cannot discuss here the more fundamental question of the appropriateness of four-dimensional space-time and of the natural constants (such as the velocity of light or Planck's constant) to the definition of life, especially of intelligent life (the so-called 'anthropic principle').

That life is a spontaneous, universal phenomenon is its primary property [48].

High Order (Low Entropy) and Life

Life is material, in the sense that it is the object of scientific observation, experiments and theoretical considerations. The differentia specifica of a living object against the background of non-living objects is its specifically high-quality order. We are speaking here about living and non-living objects which come into being without the intervention of the deliberate action of an intelligent being. The latter case cannot be discussed in this chapter because even the definition of intelligent beings has not yet been generally approved.

The high-quality order of a living object has a very small probability of emerging from non-living object having relatively low order. Of course, one must distinguish between disorder, trivial order or high order.

The quality of order has a correspondence in an area of physics which is independent of the description of matter—thermodynamics. Higher order corresponds to lower entropy. A living object has a significantly lower specific entropy than a non-living object.

That life is highly ordered (with low entropy) is its second property [49].

Life and the Inflow of Free Energy

An object with a lower entropy than the entropy of its environment passes spontaneously into the state of higher entropy; that is, its disorder increases spontaneously. Only an inflow of free (high ordered, low entropy) energy could restore the high order and remove the disorder, in the form of an outflow of bonded (high entropy, disordered) energy. The more-or-less constant flow of free energy from the environment to a living object and the removal of bonded energy is an inherent property of every living object. The nature of the source of free energy is neglected here, in spite of the fact that it is of the greatest importance.

The inflow of free energy and 'fresh' matter, and the outflow of bonded energy and waste matter, is an inherent property of every living object. This is the third property of life.

Life and the Regulation of the Flow of Matter and Energy

Living objects, with high order and with the inflow of free energy, (low entropy energy) and the outflow of bonded energy (high entropy energy), must have a boundary system to divide them from the low-order environment and a system to regulate the rate and kind of inflow of free energy and 'fresh matter' from the environment, and the rate and kind of outflow of bonded energy and 'waste matter' to the environment (Fig. 20).

The compartmentalisation of living objects and a central system for the inflow and outflow of energy and matter is the fourth property of life.

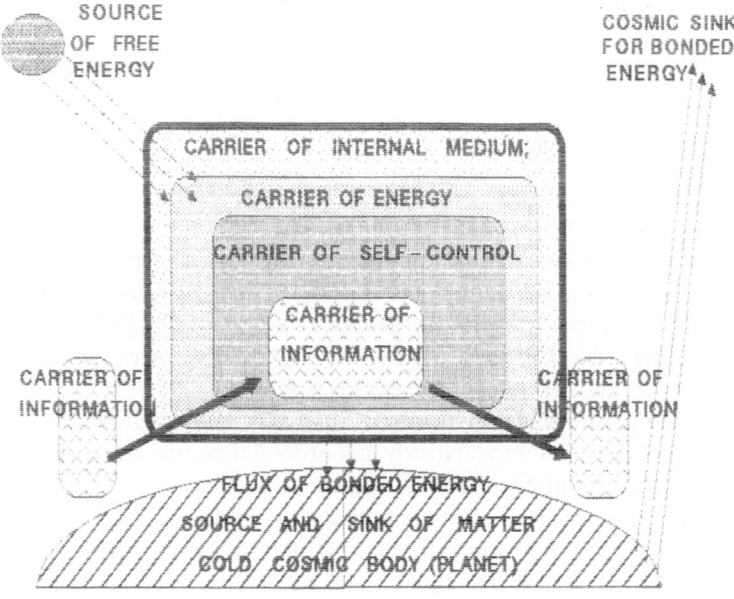

Fig. 20. General scheme of living beings

Reproducibility; Stability and Changeability

A living object cannot emerge accidentally from its less-ordered environment. It must be constructed step by step by another living object, which carried the information of its specific structure and function, even if the latter is relatively simple and small to begin with. This information must be stable enough to allow the reproduction of a living object in the first place. At the same time, the 'design' information must be slightly changed, at least for a small number of individuals to be adapted to a changing environment. The spontaneously evolving environment influences the carrier of the information and plays the role of a sieve for its fitness to survive. This causes the emergence of modified structures of living objects, and even new kinds of living systems. All the properties mentioned correspond to the statement that life as a phenomenon must be long-lived, perhaps to an age that is comparable with the age of the Universe itself. Longevity belongs not to individual living objects but to the mutually connected system of living things. The ability of a system to reproduce and exist (not as individuals) over a long period is the fifth property of living things.

To recap, a living object must have the following properties:
1) universality and spontaneity
2) high order (low entropy)
3) allow inflow of free energy and outflow of bonded energy
4) contain compartmentalisation and regulation systems for the inflow and outflow of energy and matter
5) be a carrier of information about its structure and functions, and be long-lived.

Material and Energetical Carriers of Living Systems

General Remarks About the Carrier of a Living System

Rightly or wrongly, the proposed definition of living systems is free from the specific properties of the living object itself, of the environment, of the kind of energy and kind of matter, even of the kind and amount of information needed for the reproduction of the living system. The only premise is that each living system is a material object and that energy is coupled with it. It has also been claimed that information is carried by a material carrier, without defining the kind of matter.

However, let us now translate the very general statements from thermodynamics, even if not adequately developed into the language of the branch of physics which deals with elementary particles and elementary forces, and with real matter, energy and information.

Which Elementary Forces Could be Appropriate?

The principle question to be answered is: which of the four elementary forces is the most appropriate for playing the role of carrying force for the construction of living things? In spite of the fact that the question has been formulated in a rather general form, the answer is simple and unequivocal: it is the electromagnetic force. This answer is based on the unique property of the electromagnetic force of acting both attractively and repulsively (see Table 1). The reason is obvious. If living objects must have a high-ordered, non-trivial structure, they must be controlled by attractive and repulsive actions, which are equal in strength. A trivially ordered system controlled only by attraction is, for example, a homogeneous sphere, which is too simple to carry the functions of life. The answer to our question is simple and trivial. The only elementary force which is capable of playing the role of life-carrying force is the electromagnetic one.

Why Chemical Forces?

The electromagnetic force appears in a large number of more specific phenomena, such as: (see Fig. 1).
—electrical
—magnetic
—optical
—thermal
—mechanical
—solid, liquid, gaseous and plasma state
—chemical

This means that it is not enough to select the electromagnetical force as the candidate for being the carrier of life. A more specific selection must be made. The reasons that our 'choice' are the chemical phenomena are numerous. A very simple answer could be that cosmic matter, such as gas-dust clouds, comets, meteorites, planets and their atmospheres and hydrospheres, are the scene of numerous chemical phenomena, which occur spontaneously, resulting in the emergence of rather complex and well-ordered molecules.

A more sophisticated answer relies on the fact that chemical interaction is, by its very nature, able to construct complex and well-ordered molecules, the smallest stable natural construction elements. Arguments for this statement are too numerous to be discussed here.

Each living object in the Universe is ruled by chemical interaction.

Why is Hydrogen one of the Carriers of Life?

The first property of living objects is universality and spontaneity. What kind of particles could play the role of being the carrier of the properties which are specific for each living object in the Universe, assuming that the controlling force is the chemical force and that the 'bricks' must be relatively stable? Among many arguments let us list the following properties of the material carrier of life which is being sought:

—it must be stable enough (stability of the order of magnitude of the stability of the Universe) and cosmically abundant, because the probability of the spontaneous emergence of life is, by itself, very small and cannot be reduced by using rare, exotic 'bricks',

—it must be electrically charged to give the electromagnetic force the opportunity to act. More precisely it must be an atom, because this type of particle is the object of chemical phenomena, which has been chosen as the only appropriate carrier of life,

—it should obviously be capable of interacting with all possible subspecies of chemical forces and bonds such as ionic, covalent, hydrogen-bridge and Van der Waals bonds.

The result is obvious. Only hydrogen fulfil all these conditions, especially the possibility of the hydrogen-bridge bond. Thus this specific property of hydrogen makes this element the unique carrier of life, not only here but elsewhere in the Universe:

—hydrogen is the most abundant element. Its cosmic atomic abundance is about one thousand times larger than all other elements put together, excluding helium, which is a noble gas and therefore not able to take part in chemical reactions,

—from the point of view of nucleosynthesis, hydrogen is the simplest, oldest and first element, being a substrate for the synthesis of all other elements,

—from the point of view of cosmic energetics, the burning of hydrogen is the most abundant cosmic source of free energy, in addition to the gravitational force,

—hydrogen is involved in all kinds of chemical interactions,

—hydrogen is the only element which is itself a carrier of the chemical bond. In all other elements the carriers are the electrons, at least to a first approximation,

—the number of compounds which contain hydrogen is larger than that of all other elements, including carbon,

—hydrogen compounds are very volatile, that is, having low boiling point.

It must be clear that the choice of hydrogen as the primary carrier of life and not another element, for example carbon, results in numerous significant consequences for our further argument.

Why is Oxygen the Second Elementary Carrier of Life?

The second attribute of life is a high idiosyncratic order, significantly higher than that of the non-living environment. From the above-mentioned universality and spontaneity of life, we demonstrated the unique role of hydrogen as the prime carrier of life. When we look at the second attribute, order, it must be clear that we are looking for another element which could, together with hydrogen, fulfil both criteria of universality and order. The search for such a partner for hydrogen is rather obvious. It is oxygen. Here are some arguments why oxygen is the only appropriate partner for hydrogen:

—oxygen is the second element in the list of cosmic abundance, taking into account that the real second element, helium, is a chemically non-active noble gas (see Table 12),

—oxygen-16 is the double magic nuclide (number of protons, 8, number of neutrons, 8). It is therefore very stable and abundant (see Table 9),

—it cannot be overestimated that in some main sequence stars (see Fig. 6) the source of energy is 'catalytic hydrogen-burning', and that this type of star is one of the most appropriate for playing the role of energy source for the emergence and existence of life on a neighbouring planet.

Some significant properties of hydrogen oxide—water—influence the phenomenon of life more than any other compound:

— water is the most abundant compound in interstellar gas-dust clouds.

—hydrogen oxide is the most stable hydrogen compound; that is, with the maximum Gibbs free energy of formation among all hydrogen compounds. This allows it to exist in the radiation field of medium mass main sequence stars, assuming some specific conditions, such as a protective layer of one of the products of radiolytic destruction of water, such as ozone,

—the products of the radiolysis of water have the ability to recombine very rapidly with the help of a relatively small quantity of activation energy. Other hydrides, such as CH_4 and NH_3, do not have this remarkable property. The fast and easy recombination of H and O guarantees the stability of a planetary hydrosphere, which is a conditio sine qua non of the emergence of life,

—water has a quasicrystalline property at a temperature above the melting point. This spontaneously ordered space structure plays a role in the phenomenon of life. The quasicrystalline structure results from the numerous hydrogen-bridges between the oxygen and hydrogen atoms of neighbouring molecules. For the same reason, hydrogen oxide has the almost unique property that the solid form under moderate pressure is less dense than the liquid state. Thus ice floats on water and does not sink. The importance of this property cannot be overestimated, when the stability of climate on a planet covered mostly by liquid water is considered,

—hydrogen oxide has some unique (or at least very rare) physical properties, such as an extremely wide range of stability in the liquid state, that is, the difference between the melting point and the boiling point under not too low a pressure, in a gravitational field which is not too weak. It also has a moderate dielectric constant, which allows a number of polar and ionic substances to dissolve in it, and even low polar or even unpolar substances.

Before we go further, the question of the state of the substance carrying the properties of life must be discussed. This is an old question. Can one imagine a living being which is composed only of gaseous substances? Or only of solid material? Or of liquid and quasisolid macromolecular and partially solid, crystalline substances?

One of the most general arguments is the following. A living thing needs a rather significant inflow of 'fresh' material and the outflow of 'waste' material. In a gaseous phase the velocity of material flow is highest, but the specific density is low. In the solid state the specific density is highest, but the transport of materials is extremely slow. Only in the liquid phase are both the specific density and the rate of transportation at the appropriate levels.

Why is Carbon the Third Life-Carrying Element?

The third property of life is the inflow of free energy and fresh matter, and the outflow of bonded energy and of waste matter. In searching for the appropriate chemical partner to fulfil this condition, we have some preconditions. We know that the chemical carrier of life is water. We also know that hydrogen oxide has the maximum energy of formation. One of the possibilities of incorporating some of the energy of the H_2O molecule could be to find a molecule with a general formula: H_2XO (where X is the element being sought), which evidently must be stable and soluble in an aqueous medium. One further premise is the high cosmic abundance of the element X.

The fourth property of the life is the ability of being compartmentalised, that is, partially isolated from the low-ordered environment. This corresponds to the need for a stable (but not solid) barrier of low permeability. The element X which is the object of our search must have, among other properties the ability to be a component of the stable wall of the compartment, being constructed of quasi-solid macromolecules.

All these conditions can only be fulfilled by carbon. The arguments for the role of carbon as the third elementary carrier of life are:

—carbon-12, the most abundant carbon isotope, is a product of helium-4 burning (see Table 8),

—from the point of view of the ability to carry life, carbon is the unique element able to take part in double-bond interactions (due to its small radius) with oxygen and other elements. The most important ability is its propensity for building relatively long chains with other carbon atoms (among others, resulting from its valency of 4), in a well ordered way. The intrinsic ability of carbon, amongst other elements, to produce high-ordered molecules, the smallest carriers of high order, is the reason why it is generally accepted that life in the whole Universe is intimately bound to the phenomenon of life,

—the compound H_2CO (see above H_2XO), formaldehyde, is a carrier of stored energy, is stable and is soluble in an aqueous medium. This compound has been found in polymerized form in Halley's comet. As a more general statement we can say that the appropriate molecules corresponding to the formula $H_xO_yC_z$ or its polymers, have the necessary properties for playing the postulated role: stability in an aqueous medium, solubility at least in monomeric form, carrying stored energy, and so on.

Why is Nitrogen the Fourth Elementary Carrier of Life?

The fourth property of life is the ability to control the rate of flow of energy and matter, and the ability to build the barrier wall isolating living beings from their non-living environment. From our previous reasoning we know that the molecular carriers of life have a formula $H_xO_yC_z$. Thus including the next element gives a molecule of the general type $H_xO_yC_zX$ (where X is the next element being sought) or its polymer. The required element must, of course, be stable, available in relatively high quantities, able to react with molecules containing H, O and C, dissolvable in an aqueous medium, and, last but not least, able to control the rate of exchange of energy and matter between a living being and its non-living environment. This property is called catalytic ability. Another required property of such a molecule is the ability to form quasisolid membranes (isolation walls), that is, polymeric macromolecules, assuming that the monomers must be dissolved in an aqueous medium.

The next most abundant element after H, (He), O, and C is nitrogen (see Table 12). The choice of nitrogen as the appropriate component with the required properties is not fully exhausted. However, other elements can also play an essential role. Without repeating some of the above arguments, we must say that sulphur can also play a significant role in this class of molecular carriers of life. Sulphur, in spite of its relatively high cosmic abundance (see Table 12), is not the next most abundant element, but fulfils all other criteria.

Phosphorus or Another Element as the Fifth Carrier of Life?

Now we are looking for the element which will fulfil the fifth property of life. The number of arguments available in the search for the element X, which can form a molecule with the general formula $H_xO_yC_zN_qX$ (evidently as a macromolecule), and the strength of these arguments, is weaker and less convincing. We will try to circumvent these arguments and claim that the required element is phosphorus, though stressing at the onset that phosphorus does play this role on this planet. Perhaps in another geochemical environment another element could play the same role. This could be sulphur, which would mean that instead of phosphates we would have sulphates (Fig. 21).

The role of phosphorus is a special one:

—it is possible that this is a terrestrial, local phenomenon and one can imagine that on other planets there exists a biosystem which relies on another element playing the same role of genetic carrier (DNA) as phosphorus does in the terresterial biosystem,

—the amount of phosphorus as a nutrient (structural material carrier) has been the decisive component in the increase of the global mass of the biosphere, from the early beginning. However, on the other hand, the limited amount of phosphorus was the first barrier to proliferation and evoked the 'fight for survival'.

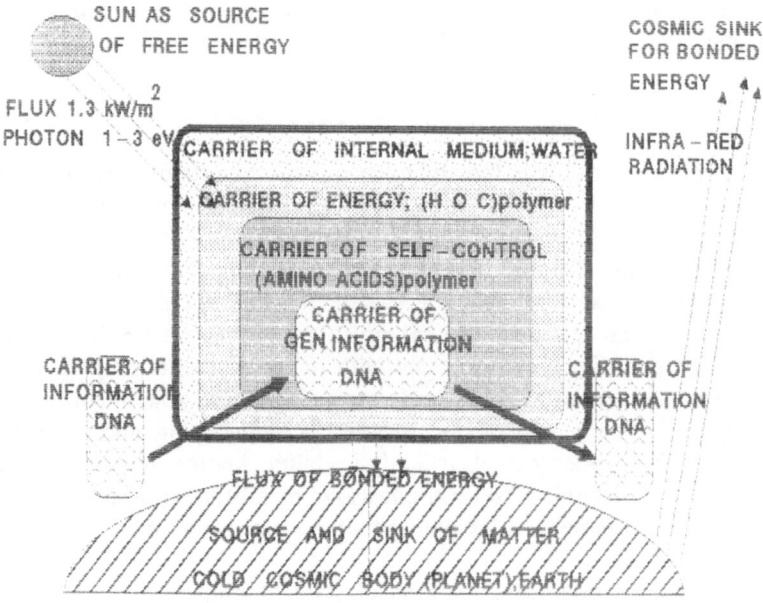

Fig. 21. Scheme of terrestrial living objects

Other Elements in Living Systems

The paramount role of water as the unique medium for life, here and elsewhere in the Universe, seems to be well acknowledged. Water in its pure state is an electrically nonconducting medium. The processes of life are of chemical, and therefore also electrical nature. The electrical conductance of an aqueous solution, as well as its osmotic pressure, result mostly from the solubility of ionic compounds, salts. As candidates for this role are the extremely electronegative and extremely electropositive (that is with a minimum of electronegativity) elements (see Table 10) but only those which have high cosmic abundance. The latter condition eliminates Li, Be and F. The most suitable elements are the electropositive Na, K, Mg, and Ca, in the form of cations, and the electronegative Cl, in the form of anions. Because the previous reasoning concerning the main carrier of life included C, S and P, therefore in aqueous solution the complex anions of carbonates, sulphates and phosphates, for example, must also be considered.

Other metallic elements may also play decisive and unique roles as active centres of the organic catalysts, especially the transition elements with 3d electron configurations such as V, Cr, Mn, Fe, Ni, Cu and Zn. These elements lie on the neighbourhood of iron, whose atomic nucleus (Fe-56) is the most stable of all atomic nuclei and therefore the very probable product of numerous nuclear reactions, both of fusion and fission (Table 17).

Table 17. Elementary composition of the terrestrial biosphere

Atomic number	Element	Composition of biomass in the living state, percentage	
		Atomic	In weight
Carriers of life			
1	H	62.9	10
8	O	24.8	63
6	C	10.5	20
7	N	1.1	2.5
15	P	0.2	1.1
16	S	0.03	0.14
Important elements			
11	Na	0.02	0.1
12	Mg	0.02	0.1
17	Cl	0.03	0.16
19	K	0.02	0.11
20	Ca	0.4	2.45
26	Fe	0.015	0.01
Less abundant elements			
9	F		0.004
14	Si		0.03
23	V		
24	Cr		
25	Mn		0.03
27	Co		
29	Cu		0.0002
30	Zn		0.0001
34	Se		
35	Br		0.0003
50	Sn		
53	I		0.0001
	Total	100.00 at%	100.00 w%

Thermal Conditions for the Existence of Life

Suppose that living systems exist in relative equilibrium only when random thermal motion breaks the intermolecular bonds not more than 1 time in a million movements. Under these more or less arbitrarily estimated conditions, the hydrogen bond, which has an energy of about 0.03 electronvolts (1 eV/bond corresponds to 96.5 kilojoule/mole, and $E = k \cdot T$, where k = Boltzmann coefficient, $k = 8.15 \cdot 10^{-5}$ eV/K, T = temperature in Kelvin, E = energy in eV/molecule), can only exist when the environmental temperature is not higher than approximately 40 K. For hydrogen bonds with energy of 0.1 eV/molecule, the maximum temperature equals 400 K, and for bonds of 2 eV, just below 2000 K.

From our precondition, that life can exist only in water. The temperature of the environment is, under medium pressure, limited to the range 273 K to 380 K. The pressure of the atmosphere and the strength of the gravitational field of the planet are obviously mutually related. On a planet of medium size, the appropriate distance from the central star, and with the requisite thermal balance, there is some chance of preserving hydrogen oxide in its liquid form at all times, at least on a part of the planet's surface (see Table 15). Earlier studies have been relatively pessimistic about the probability of the existence of a hydrosphere. Recent studies, however, are much more optimistic and significantly increase the probability for the emergence of life in the Universe.

The most important condition for the existence of life is the existence of very long-lived sources of free energy. In a naive model, one source of free energy can be in the form of stored chemical substances which can react with other substances, also existing in the neighbouring environment. Both substances must be present in such amounts that the system of living things could evolve over a period of gigayears. It is simple to show that such storage of chemical substances on the planetary scale is impossible.

Only a neighbouring main sequence star can play the role of the source of free energy. It is worth repeating that the Earth today receives an energy flux of $1.75 \cdot 10^{17}$ W, while the Sun emits approximately $4 \cdot 10^{26}$ W. The Earth profits from only $4.3 \cdot 10^{-10}$ of the total solar radiation.

Not only the amount of the stellar radiation, but also the radiation spectrum, are of significance. If life is always a chemical reaction, photons must carry energy corresponding to the energy of the chemical bond, that is about 1 eV, but not higher then 2.5 eV, because this would cause disruption of the water molecules.

Biosphere in the Past

The First Living Structures: Protobionts and Eobionts

The so-called 'primordial soup' contained a rather concentrated aqueous solution of substances which are very easy to synthesize. (Caveat: Why is there no evidence of this primordial soup in sediments?) Among these substances are probably NH_3, H_2S, CO_2, H_2CO, HCN, amino acids (at least the same as in the Murchison meteorite, see Table 14) simple pentoses and hexoses, phosphoric acid, and salts of Fe-II and other elements such as Mg, Ca, K, Na, etc. Hydrogen cyanide polymerized and, after hydrolysis produced some amino acids and purines and pyrimidines. It has also been suggested that some polymers, for example polypeptides and simple ribonucleic acids, have also been present from the very beginning.

These macro-molecules can, from the beginning, play the role of space-organising membranes and process-rate controlling agents (catalysts, enzymes). The first structures which began to transfer information about their own internal arrangement and most important functions were formed spontaneously by a random process. This assumption has a higher probability if it is postulated that not 20 amino acids were present, as in modern living beings but only a few amino acids and appropriately simpler codons, for example only due to a pair of guanine and cytosine. Instead DNA, RNA was probably the carrier of genetic information, especially as RNA shows true catalytic, enzymatic properties. If a molecule can catalyse its own reproduction then the 'chicken-and-egg' problem seems to be

solved. Spontaneous polymerization of charged amino acids could be an important step in this process. These first highly structured organisms are called here
'protobionts'.

Protobionts were able to perform all of these processes necessary for their
existence and reproduction, assuming that all the construction material required
was floating in their direct neighbourhood in the primordial soup. Free energy
could be carried by ATP (adenosine triphosphate), which must be a product of
spontaneous synthesis in the primordial soup due to the action of ultraviolet
radiation, which could reach the Earth's surface because the atmosphere lacked
free oxygen and ozone.

Several environments have been proposed as candidates for abiogenesis. One of
these is an evaporating lake, lagoon or shallow marine area. A second possibility is
a hydrothermal spring (at the present time found on mid-ocean ridges) having a
high concentration of potential catalysts. A third possibility is the soil or a shallow
subsurface containing clay. In all of these environments, the time-scale for abio
genesis seems to be only of the order of some megayears.

The next big jump in the evolutionary process is caused by the exhausting of the
ATP in the primordial soup. The next generation of living beings which could
carry this name are undoubtedly the 'eobionts'. These are capable of producing
ATP by means of fermentation, relying only on substances in the primordial soup.
These organisms must be classified as chemolithotrophic ('chemoautotrophes').
Some eobionts consume other eobionts, perhaps smaller ones or those with weaker
outer membranes. These eobionts are the beginning of 'heterotrophic' organisms.

Environment and the Biosphere in the Past

There are well-founded opinions that one of the principal properties of life is the
intrinsic one of proliferating exponentially to a limit ultimately set by the availability of critical resources. The limitations of the most important primary energy
flux—solar radiation—is in most cases not the bottleneck for proliferation. In
most cases it is the notable limitation nutrients such as phosphorus, nitrogen, and,
to a lesser degree, iron.

The young Earth could well have been in such a state of global biotic saturation,
with profilic microbial ecosystems monopolizing the totality of suitable habitats.
The evidence for this can be seen in the variety of surface-dwelling (benthic) species
(largely procaryotic) which have been well preserved in the form of impressive
fossilized stromatolite carpets since early Archean times.

It can thus be assumed that autotrophy, particularly photoautotrophy, has been
extant on Earth, both as a biochemical process and as a geochemical agent, for at
least 3.8 Gyr. The biogenicity of 'stomatolites' (procaryotic microbenthos) is
generally accepted. However, the biogenicity of the presumably older cellular
microfossils is still disputed [51].

It is not difficult to imagine ancient habitats saturated with microorganisms to a
degree approaching that of man-made eutrophicated environments of the present
world. As the Precambrian continents were virtually lifeless, there were no land-
living autotrophic beings to retain rare nutrients such as phosphorus. The young

ocean was probably eutrophicated by a factor of at least two compared to the present ocean. In fact, accepting the role of phosphorus as the ultimate determinant for the size of the global biomass implies the existence of a state of global biotic saturation or plenitude from the time when autotrophic living beings first appeared on the Earth.

Further Evolution of the Biosphere

The day comes, however, when the primordial soup is so depleted that no more molecules are present which can be used in the free-energy releasing fermentation process. This is the next starvation barrier on the way to the evolution of a terrestrial biosphere. The next step is to manage to use solar light in the form of one photon for releasing energy-carrying hydrogen from hydrogen-carrying molecules, such as H_2S and CH_4. In these molecules, hydrogen is relatively weakly bonded, which means that they are not very stable. They would also not have been relatively very abundant on the early Earth.

The most abundant hydrogen-containing molecules were, of course, those of water. However, they are so strongly bonded that only an ultraviolet photon is able to release the hydrogen. Ultraviolet destroys water molecules, and therefore also intracellular water, and kills living things. The direct utilisation of UV by living organisms must, therefore, not take place in the organism if it is to survive. Another way must be found. The two-photon system (photosystems I and II), in which photon energy can be stored for a long enough time in exciting molecules, can be used for the photolysis of water. Hydro-photolytic systems are the basis of very 'autotrophic' organisms. Nitrogen fixation by various unicellular organisms must have appeared quite early on, because of the exhaustion of sources of NH_3 or nitrogen containing substances.

A crucial question, which still has not been solved, is that concerning the origin of photosynthetic microorganisms, more exactly of chemophototrophic microorganisms. Recently some investigators [51] have claimed, based on the isotopic composition of carbon in some sediments, that photosynthesis must have existed as a biochemical processes for almost 4 Gyr, primarily operated by microbial ecosystems. At the same time these biochemical processes were important agents in the geochemical transformation of the Earth's surface (Table 18).

Further evolution then proceeds in classical biotic form, which advances through prokaryotic, and later through eukaryotic, unicellular organisms (both levels in the form of chemotrophs, chemophototrophs, phototrophs and heterotrophs). The eukaryotes evolve over some 750 Myr into multicellular organisms (only in the form of phototrophs and heterotrophs). Then roughly 580 Myr ago these multicellular organisms evolve exoskeletons (calcareous and siliceous) and, later, endoskeletons. The continents can be conquered.

The emergence of the nervous system, the central nervous system and then the brain allows the terrestrial biosphere to reach the highest level of evolution. The human brain is capable of initiating a new kind of trophical level. Humans, being typical heterotrophes, can achieve such a high level of technology, including the synthesis of almost all nutrients, inclusive of proteins, vitamins, etc., that it seems

Table 18. Scheme of kingdoms and energy sources

			Organization levels		
			Prokaryotic	Eukaryotic	
			Unicellular		Multicellular
Energy source	Autotrophy	Chemotrophy	Yes	Yes	No
		Chemophototrophy	Yes	Yes	No
		Phototrophy	Yes	Yes	Green plants
	Heterotrophy	Absorption	Yes	Yes	Fungi
		Digestion	Yes	Yes	Animals

to be possible to foresee the emergence of 'technotrophes'. A real new step in the evolution of human beings is the emergence of self-consciousness which took place probably over some tens of thousands of years, together with the 'invention' of speech. Only for a short time have human beings begun to consider the relationship between themselves and their natural environment. Both of these factors—self-consciousness (knowledge of one's internal environment) and consciousness of responsibility for the natural environment—are the most significant markers of our time.

Extinctions and Discoveries

The evolution of terrestrial life has been influenced by several periods of large extinctions of continental and marine species, disappearing in a relatively short time. The five most pronounced mass extinctions in the marine environment came in the late Ordovician, the middle-late Devonian, the late Permian, the late Triassic, and the late Cretaceous periods. In the continental environment some mass extinctions occurred at the same time, but not always of the same magnitude [55, 56, 57, 58, 59].

The extinction of species is often rather similar to the decay of radioactive nuclides. This means that at successive time intervals the proportion of species that becomes extinct is essentially constant. The typical order of magnitude of the 'half-life' is some millions of years. Major exceptions to this pattern occur as a consequence of mass extinction events, a type of catastrophe. In these cases, the half-life is of some thousands of years, but there is no evidence for a species half-life of years or months.

The hypothesis has been put forward that in the past, especially at the end of the Cretaceous period (the Cretaceous/Tertiary boundary), about 65 Myr ago, the fall of a large meteorite (it could have been a asteroid, or perhaps a comet) with a diameter of 7–10 km and mass of 10^{14} kg, released energy of around 10^{23} J. This corresponds to the terrestrially absorbed solar radiation during a period of some days. The fall of a meteorite mostly acts through the impact of its kinetic energy, dissipated in an extremely short period of time and extremely localized. The dust aerosol generated by the impact has been estimated to be 5 Eg ($5 \cdot 10^{12}$ ton), spread uniformly around the globe for a period of several months, resulting in subfreezing temperatures over all land masses and a global loss of photosynthesis for up to one year [42].

Some scientists believe that the arrival of cometary substances such as cyanide from the nucleus could provide the poison to kill oceanic planktons. However, it is more probable that cometary cyanide would have been decomposed and therefore detoxified by the intense shock-wave pressure following the comet's impact with the Earth's surface.

Catastrophes Caused by Terrestrial Events

There exist at the present time approximately 500 active volcanoes, only a small fraction of which are of the type that produces explosive volcanic eruptions. It has been claimed that, if 1/5 are explosive and all of these erupt during a century, the mean temperature of the Earth would only be 1 °C cooler than normal. Increased volcanism over an extended period, associated with lava flows with volumes greater than 100 cubic kilometres, could inject large quantities of sulphate aerosols into the lower stratosphere. This could lead to the production of immense amounts of acid rain, the reduction in alkalinity and pH of the surface ocean, global atmospheric cooling, and ozone layer depletion.

For the sake of comparison, let us look at some data concerning the largest volcanic eruption in recent times. On August 26, 1883, in the Sunda Strait occurred the explosion of Krakatoa. This catastrophic event released energy equalling 1 EJ (10^{18} J) and ejected material totalling 10 cubic km, or $50 \cdot 10^{12}$ kg. Some of this material, in the form of ash, was propelled to a height of 80 km, even higher than the ozone layers. It is clear that a large number of Krakatoa-sized eruptions are needed to produce large temperature changes. In around 1500 B.C. in the Aegean Sea, the volcano Santorini released an energy of 10^{21} J. Another volcanic eruption at Tambora in 1815 with the ejection of 100 to 300 cubic km of material, containing 0.15 Pg of sulphate, was followed by the legendary 'year without a summer'. The average volcano-induced cooling of the northern hemisphere in the summer of 1816 was only 1–2 °C.

An earthquake having a magnitude of 8 on the Richter scale, which is large but occurs not infrequently, corresponds to an energy release of $50 \cdot 10^{15}$ J. This is the same amount of energy as that released by the fall of the Tunguska object. Perhaps it is not strictly fair, but it seems reasonable to make a comparison between these natural catastrophes and the impact of a nuclear war. In general, the reference case corresponds to a nuclear war with the use by both sides of 6500 Megaton nuclear bombs. Roughly, this equals the release of 65 EJ ($65 \cdot 10^{18}$ Joule) of energy. Including secondary effects, such as forest and city fires, this results in the release to the atmosphere of 360 Tg (360 million tons) of smoke, containing about 65–70 Tg of soot. It must be stressed that volcanic aerosols, in general make poor analogues for the effect of smoke from nuclear war, because the mechanisms of injection into the atmosphere, the morphology and chemical properties, and the optical properties of carbonaceous smoke versus volcanic ash or sulphuric acid droplets are different. According to recent studies, the maximum summertime, northern hemisphere, average land surface temperature would change by 5–15 °C; similar to the normal mid-latitude change from summer to autumn. There are other possible mechanisms, however, which could also significantly change the global climate; for example, failure of the normal Asian summer monsoon or dramatic stratospheric ozone reduction.

It must be stressed that the impact of a large cosmic body may have another effect, from the plume of vaporized rock and water which would have risen beyond the stratosphere. The air would have been compressed and heated to temperatures of 2000 K, hot enough so that the nitrogen in it would have reacted with the oxygen. Within hours the atmosphere would have received 10^{12} kg of NO and NO_2. And within days the rain clouds would have filled with nitrous and nitric acid. Local rainfall would have a pH of 0 to 1. Some 2 years later, after the nitrogen oxide cloud has been thoroughly mixed, global rainfall would have been diluted to a relatively benign pH of 4. The impact can thus cause a smog of nitrogen oxides and waters which have been poisoned by trace metals (Be, Al, Hg, Pb and others) leached from soil and rocks. The chemical effects of the impact are therefore at least as important as the better-known physical effects, such as dust and smoke generation, and the subsequent cooling of the global climate. Plants, which could have survived in the form of seeds and roots, would be relatively unscathed, but the largest animals, the dinosaurs, which had had to endure both starvation and asphyxiation, would be wiped out.

The Biosphere Today and in the Future

Biomass and the Production at Present

The biosphere is, to a first approximation, a steady-state system. The total mass of the present biosphere, given in so-called dry mass form, equals 1841 Pg ($Pg = 10^{15}$ grams). Of course, this number and all later mentioned data are uncertain and differ rather significantly in different publications. The continental biomass equals 1837 Pg, which is 99.78 percent of the global biomass. The marine biomass equals 4 Pg, which corresponds to only 0.22 percent of the global biomass. From this point of view the oceans seem to be a lifeless desert. Recently, new measurement techniques, including satellite observation, seem to show a significant increase in the value of the marine biosphere, not by a factor of two, but by as much as an order of magnitude.

This is not so from the point of view of the annual productivity of both types of biosphere, continental and marine. The annual productivity of the continental biosphere equals around 132 Pg and that of the marine around 100 Pg. Table 19 shows these and other values for the productivities of the continental and marine biospheres given by other authors. For each kilogram of biomass produced by the marine biosphere the continental biosphere produces about 1.3 kg. Even if we take into account the fact that the surface of the oceans is about 2.4 times larger than

Table 19. Mass and production of the biosphere today. 1 kg dry organic matter $= 0.455$ kg carbon, corresponding to $20\ MJ\ Mm^2 = 10^{12}\ m^2 = 1$ million square kilometres; $Pg = 10^{15}$ grams $= 10^{12}$ kg $= 1$ Gigaton

Region	Surface Mm²	%	Biomasse Pg	%	Production Pg/year	%
			Continents			
			Continental Climatic Zones			
Polar	8.05	1.6	10.5	0.57	0.5	
Boreal	23.2	4.5	355	18.2	11	
Subboreal	22.5	4.5	213.5	11.6	12	
Subtropic	24.2	4.8	248.5	13.5	26	
Tropic	55.8	10.8	1029	55.9	82	
Glaciers	13.9	2.7	0.0	0.0	0	
Lakes, streams	2.0	0.4	0.2	0.1	1	
Total	149.3	29.3	1837	99.78	132	57
			Continental Bionic Zones			
Forests	31	6.1	1312		48.7	21
Woodland						
Grassland and						
Savannah	37	7.3	462	25.1	52.1	22
Cultiv. land	16	3.1	15	0.8	15	6.5
Arctic, alpine	25	4.9			2.1	1
Deserts	30	5.9			3.1	1.3
Wetlands	6	1.2			10.7	4.6
Lakes, streams	2	0.4			0.8	0.3
			Oceans			
Total	361	70.7	4.0	0.22	100	43
			World			
Total	510.3	100.0	1841	100.0	232	100

Annual production Pg/year					
	This book	(10)	(60)	(63)	(66)
Ocean	100	55	55	92	110
Continents	132	117	141	132	–

that of continents, the total annual marine production does not allow the ocean to be called a desertlike area (Fig. 22).

Productivity, expressed in kg of dry matter produced per square metre and per year is for the global biosphere 0.445 kg, for the continents 0.886 kg, and for the oceans 0.227 kg. The specific productivity (that is the annual productivity per unit of biomass) equals approximately 25 kg per kg and year for the oceans, and 0.072 kg per kg and year for the continents. These, in some way astonishing values result from the different lifetimes of species members of the both type of biosphere, that is the average residence time. The marine species, on average, live approximately 15 days, the continental approximately 14 years. The marine biosphere consist primarily of unicellular short-lived organisms (algae), the continental of long-lived large trees.

A rather significant problem concerning the productivity of the global biosphere results from the possible increase of atmospheric carbon dioxide. It is not yet clear if the response of the biosphere will be positive, that is, the productivity will increase with carbon dioxide loading and at what rate and how differentiated in different regions and climatic zones [61]. It is obvious that the repercussions of this phenomenon are of the greatest importance for further global decisions concerning the use of fossil fuels (Fig. 22, 23 and Table 19).

The whole problem of the coevolution of global climate and global biosphere, especially on the regional scale, seems to be inadequately researched [62].

Direct and Indirect Use of Biomass by Humans

For the first time in the history of this planet, one species alone out of at least 2 million, and possibly more than 5 million, presently existing species, controls the use of the total, global primary production of the biosphere and other resources too.

Some calculations can be made concerning the present rate of indirect annual use of dry biomass by humans [63]. It must be stressed though that some of these data differ by a factor 4 or even more, according to source. In spite of these uncertainties, however, it is of the highest interest to try to formulate an estimate. The primary annual production of the biosphere includes not only cultivated land, pastures, natural grazing lands in use, and the grazing land burned, but also the

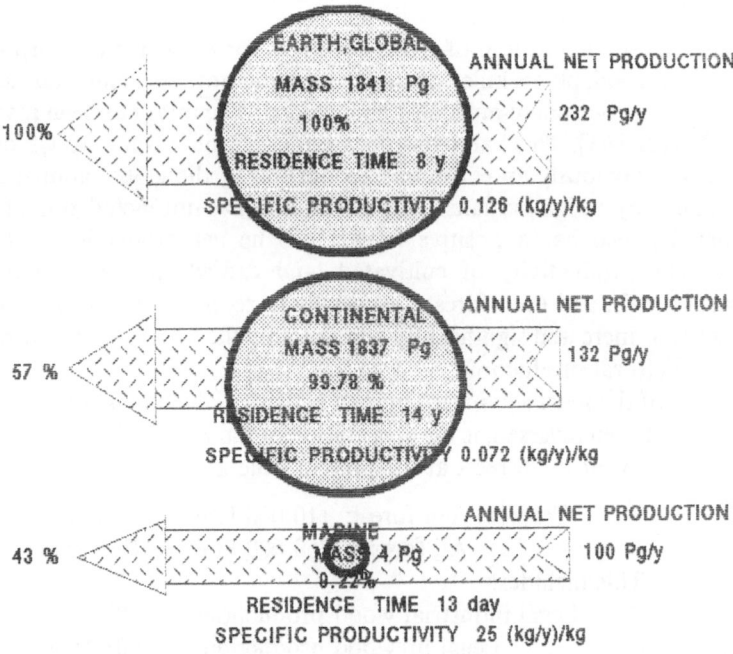

Fig. 22. Productivity of the continental and marine biospheres

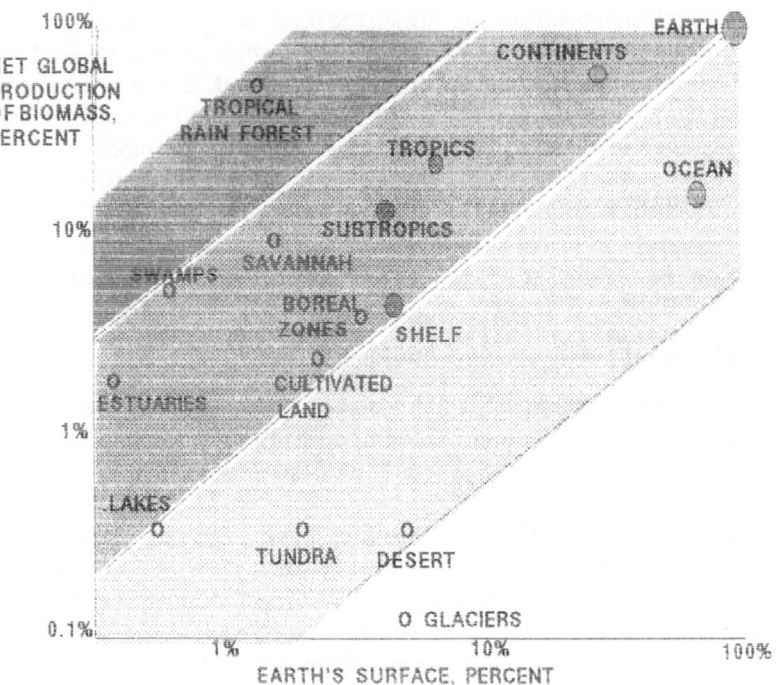

Fig. 23. Regional productivity of the biosphere

forests, such as the harvesting of wood products for construction purposes, for pulp and as firewood, plants killed during harvest but not used, and land clearance. Use of the marine biosphere must also be included. The total has been assessed to equal 40 Pg/year [63]. This value has been criticized, but even only 1/2 or 2/3 of this quantity is obviously an enormous proportion of the global annual production, consumed by 5 giga humans. This calculation has not included the fact that human activity also has a positive impact on the net production rate of the biosphere. The productivity of cultivated land exceeds probably that of the previous savannah, and even forests, especially in recent times when agricultural productivity has increased significantly and not only the 'best land' has been taken over for agricultural production.

Maybe one of the most significant human activities which must be halted in the near future, and even reversed, is the devastation of the world's forests. The annual rate of destruction of the forests at the present time is:

$$\text{Tropical rain forest: } 110\ 000\ \text{km}^2 = 0.11\ \text{Mm}^2$$
$$\text{All forests: } 250\ 000\ \text{km}^2 = 0.25\ \text{Mm}^2$$

This includes:

$$\text{Legal industrial wood production: } = 0.07\ \text{Mm}^2$$
$$\text{Legal firewood production: } = 0.03\ \text{Mm}^2$$

Note: Illegal destruction of forests is the most significant.

Table 20. Amount of net primary production used directly by humans and domestic animals Accord. Vitousek [63]

Source	Pg/year
Cultivated land, food	0.8
Domestic animal fodder	2.2
Wood products	
Construction and pulp	1.2
Firewood	1.0
Fisheries	2.0
Total	7.2

Productivity of the Marine Biosphere

Primary production by marine microalgae is believed to be a critical factor regulating atmospheric carbon dioxide concentration. The high specific annual productivity of the marine biosphere which has been estimated above as equal to 25 kg of dry biomass per kg of living biomass, is responsible for relatively rapid recirculation. Assessments of photosynthesis in the open oceans (more than half of the Earth's surface) and the related transport of organic carbon to the deep ocean, vary by as much as an order of magnitude [66]. Discrepancies are attributed to different temporal and spatial scales, reflected by instantaneous rate measurments, as opposed to seasonally averaged measurements. Satellite-based extrapolations of pimary production can be used to give the best estimate.

The rate of primary production of benthic prokaryotes (specifically cyano-bacteria, which are the simplest phototrophes) reaches the formidable proportions of 18 to 26 g of dry biomass per square metre per day. The current average daily marine production rate equals 0.6 g dry biomass/m² and varies from 25 g/m² to less then 0.5 g/m² in the open ocean. Microbial communities are among the most productive ecosystems on the Earth today.

Recently new data concerning the primary production of phytoplankton has been enlarged which can have enormous impact on the energy-mass balance of the marine biosphere. The well-known genus of phytoplankton includes the so-called microplankton, that is, eucaryotic green algae with a diameter of some micro-metre. Recent measurements indicate that the role of picophytoplankton of diameter between 0.2 to 1 micrometre is extremely significant. The number of these cyanobakteriae in one cubic metre equals 10 million, which may be equivalent to a mass of about 10 micrograms per cubic metre of oceanic water. In spite of this, it has been suggested that picophytoplankton could be responsible for 20 to 80 percent of primary production.

As we have seen, the recycling of phosphorus in the early ocean was essentially different from the present, because of the absence of the continental biota. This suggests a different mode of cycling of almost all important components of global material cycles. For example, phosphate cycling in the Archean ocean was essen-tially different, as the rivers carried much more phosphorus because the continent was lifeless and there were no microorganisms there which could retain phosphorus. The difference in the phosphorus cycle obviously has repercusions on the marine carbon cycle. In addition to phosphorus, iron is the essential component in the open ocean which limits the growth of plankton, the primary producer. Among other sources of iron, the dry winds of the continents contain dust incorporating iron oxides.

A simple comparison of two independent numbers reveals an interesting relationship. The amount of uranium dissolved in oceanic water equals $4 \cdot 10^{12}$ kg, corresponding to a concentration in approximately $1.3 \cdot 10^{21}$ kg of the oceanic water of only 3 ppG. For comparison, fish total only some 0.47 ppG of all water, that is, 0.47 grams of 'fish' in each million litres of water. Comparing those figures shows that uranium, one of the rarest elements on the planet, has a concentration 7 times more than the fish in the sea and equal to the total dry mass of the marine biosphere. At different times and in different areas, particularly near the surface, fish tend to concentrate in large shoals, but uranium is more or less homogeneous throughout the ocean at all times and at all depths.

Marine Food Production; Marineculture?

Two-thirds of the surface of this planet are covered by the oceans; about 361 million square kilometres. However, the oceans contain only 0.22 percent of global living things, that is, 4 Pg of dry biomass. However, we cannot presume that recent estimation of the marine biosphere based on satellite measurements will not change this value. An increase by a factor of two or even more cannot be excluded. In the meantime, we will stick to the value of 4 Pg of dry global marine biomass. Now, the productivity of the oceans is relatively high, reaching 100 Pg of dry biomass per year, that is, 43 percent of the global biogenic primary production. The productivity of the marine biosphere equals 25 kg dry biomass per year and kg of dry marine biosphere. This is approximately 350 times greater than the productivity of the continental biosphere.

Measuring in terms of unit surface area results in specific productivity of the oceans equal to 0.28 kg of dry biomass per m² and year. This is roughly a quarter of the productivity of the continental biomass. The ocean's production is not evenly distributed over its whole area. Some areas are highly productive, others very unproductive. (Table 21, 22).

The average productivity of the marine biosphere of 0.153 kg dry mass per m² and year means that only 0.08 Watt/m² of solar radiation has been effectively transformed into free energy of the biosphere. Since, the total solar flux, on average, can be assumed to be 170 W/m², this corresponds to an efficiency of 0.05%. The remaining solar radiation, about 99.95%, has not been directly involved in photosynthesis in the marine biosphere.

What is the reason for this? Why are the oceans so unproductive? The reason that the world's oceans are deserts, from the point of view of biogenic productivity, is the following. The presence of practically unlimited amounts of water and very large amount of solar radiation, such as in the tropics, is not sufficient to generate intensive photosynthesis. The most important limitations are the lack of certain

Table 21. Productivity of the marine biosphere

Parameter	Unit	Upwelling and Estuaries	Coasts	Open Ocean	Total Oceans
Oceans					
Surface	$10^{12}\,m^2$	0.36	35.0	324	361
Surface	Percent	0.1	9.9	90	100
Total marine biosphere					
Productivity	g/m^2year	1400	500	270	280
Productivity	Watt/m^2	0.90	0.25	0.16	0.18
Production	Pg/year	2	20	78	100
Number of trophic levels	Number	1.5	3	5	4.8
Fish					
Productivity	g/m^2 year	270	3.5	0.005	0.61
Production	Tg/year	96	122	1.6	221
Harvest	Tg/year				80?

Table 22. Constraints on continental and marine biogenic productivity

Parameter	Unit	Open Ocean	Continents (including desert, ice)	Ratio of ocean-to-continent
Surface layer of oceans versus continents				
Solar radiation	W/m^2	170	170	1
Water	–	unlimited	1 metre of rainfall	very large
Mineral content	kg/cub.m	35	more 100	less 1/3
Carbon dioxide	ppM in air	>350	350	more 1
Nitrogen, chemically bonded	g/cub.m	0.01	>0.04	less 1/4
Phosphorus	g/cub.m	0.02–0.03	5000	10^{-6}
Iron	g/cub.m	0.01	>5000	$<10^{-6}$
Suspended solid	particles per cub.m	limited	unlimited	very large
Productivity (annual)	g/cub.m	140	785	1/5
Deep ocean layer of ocean; >1000 m				
Nitrogen chemically bonded	g/cub.m	0.24–0.35	0.04	10?
Phosphorus	g/cub.m	0.05–0.09		
Suspended solid	particles per cub.m	?		

essential elements, such as iron and phosphorus, and probably the lack of solid particles to play the role of passive carriers of some photosynthetic micro-organisms.

The deeper layers of the ocean, below 1000 m, however, contain a higher concentration of the essential elements, such as P and Fe, so that in those areas where the cold deeper layers well up to the surface the productivity increases dramatically. In these areas the rate of production is some eight times higher than in the open oceans. Due to the rather complex pyramid of trophic levels, the productivity of the fish population itself varies from $270 \, g/m^2$ per year in these favorable areas to only $0.005 \, g/m^2$ per year in the open oceans. Of course, in the lowest trophic level the productivity of green algae in the mineral-rich cold water of the upwelling zones is relatively high at $0.2 \, kg/m^2$ per year. Of this amount, 3/4 is consumed directly by small and larger herbivores. Small crustaceans, krill (8–60 mm long), have a surprisingly high productivity of about $0.1 \, kg/m^2$ year. Fish, being on the third and fourth trophic levels, are limited in productivity. Going back to the basic relationship—to that of primary solar energy—it is found that only 5 ppM of solar radiation finds its way into the nets of man in the form of fish meat.

In spite of all this, the marine biosphere is a significant producer of human food; in energy terms approximately 2%, in total protein around 8% and in animal protein about 25%. The present production of fish and other marine animals, such as jellyfish, totals about 80 Tg/year (80 million tons/year) corresponding to around 1/3 of the global production of fish, which equals some 220 Tg/year, and has been in equilibrium for at least ten thousand years.

Present and Future Productivity of the Continental Biosphere

An increase of atmospheric carbon dioxide influences the biosphere in two ways: indirectly through its impact on the global and regional climate, including the amounts and temporal distribution of rainfall and of influx of solar radiation, and directly through an increase of concentration of CO_2 in the neighbourhood of leaves and its subsequent transport through the stomata. Rapid changes in photosynthesis and water-use efficiency, due to changes in stomatal control under enhanced CO_2, have been observed in laboratory experiments. All these effects are expected to alter forest productivity and the regional distribution of species. Forests are responsible for most of the continental biomass production. However, forest response to climatic change depends in part on changes in soil water and nitrogen availability, which limit tree growth.

Some model calclulations have been made corresponding to a doubling of atmospheric CO_2 concentration (from the present level of around 350 ppM to 700 ppM), and the resulting rise in global temperature is 2–4 °C, with greater warming at higher latitudes than near the equator. This doubling is expected to occur over the next one hundred years. Several geographically explicit vegetation models suggest profound changes in the distribution of the major biomass, particularly in norther temperate and boreal regions. Insofar as nitrogen and soil water are two of the chief limiting factors to forest growth, such as in eastern

North America, the response of tree growth is rather complex. The model calculations lead to conclusions that interactions between vegatation and water, and nitrogen availabilities, again as in North America, may result in bifurcation of the forest ecosystem response: (a) increased productivity where soil water is not limiting and nitrogen availability is enhanced, and (b) decreased productivity where both water and nitrogen become more limited [66].

Biosphere and the Use of Solar Energy

The Efficiency of Green Plants

The total net annual productivity of the global biosphere has been estimated as 232 Pg of dry mass. One kilogram of dry biomass resulting from photosynthesis contains around 20 Megajoules (some authors estimate a value of 17 MJ). From this we can calculate the annual amount of effective (net) solar energy used in photosynthesis:

$$(232 \cdot 10^{12} \, kg/y) \cdot (20 \, MJ/kg) = 4.64 \cdot 10^{21} \, J/y$$

which corresponds to an energy flux of:

$$(4.64 \cdot 10^{21} \, J/y)/(3.15 \cdot 107 \, sec/y) = 150 \, TW \, (Terawatt)$$

In this calculation only the net annual productivity has been taken into account. The gross annual productivity comes to roughly 1.5 times more than the net value, that is, around 350 Pg/y. The total solar energy flux required for photosynthesis is thus 225 TW. This amount corresponds to about 0.19 percent of the solar energy which is effectively absorbed by Earth.

Let us consider the mechanisms of primary production, that is, the process of photosynthesis in green plants, which results in such a low efficiency. As we have roughly calculated, plants only use less than 1 percent of the solar energy striking the surface on which they live. The cause can be generally characterised in the following way (in percentages of total influx of solar light):

The total solar energy flux 1.00
Only 80% of surface is covered by plant 0.80
Around 16% of solar light is reflected 0.64
Approximately 4% of light is transmitted. 0.60
10% of light is absorbed by other than leaves. 0.50
27% of solar light is not involved in photosynthesis because of
low photon energy 0.23
20% thermodynamic and metabolic losses during all steps of photo-
synthesis 0.03
1% losses due to instantaneous deficit of CO_2 . . . . 0.02
0.7% losses of respiration (gross-nett) 0.013
0.3% losses because of disease etc.. . . . . 0.01

The real efficiency of utilization of solar radiation is significantly smaller than our optimistic rough calculation. The flux of global solar energy needed for growing plants in equilibrium equals at least 22 000 TW. However, even this figure is not complete and will be significantly revised in the following chapter.

The low efficiency of utilization of solar energy means also that the remaining 99 percent of the energy flux must be transferred back to the environment. The evaporation of water is one of the most important ways of removing the bounded, low-entropy energy to the atmosphere. The transpiration ratio—the mass of water used by a plant to the accumulated dry biomass—equals from around 140 for evergreen trees, to 300 for maize, to over 800 for deciduous trees. Evaporation of, for example, 500 kg of water need more than 1 GJ of thermal energy. The 1 kg of dry biomass produced contains around 20 MJ of chemical energy. The ratio is thus 100:2.

Number of Species—Past, Present and Future

Another result of the extremely negative aspect of human activity, in addition to the quantitative overuse of the primary production of biosphere, is a negative qualitative impact: the destruction of some species. The first human impact, the quantitative, seems to be reversible, even if the price of restoration will be much higher than the current 'profit'. The second human impact, the qualitative, is irreversible, as the loss of numerous species cannot be reversed.

During the evolution of life on the Earth, probably some millions of species occurred, and the number of different multicellular organisms could have reached 10^{23}. At the present time, approximately 10^{28} cells exist, both in unicellular and multicellular organisms, which corresponds to a fresh biomass of around $(10^{28}$ cells$) \cdot (5 \cdot 10^{-13}$ kg/cell$) = 5 \cdot 10^{15}$ kg.

The two million presently identified species of living things, and probably a further three million still unidentified, are very unevenly distributed over the Earth. The tropics cover some 42 percent of the continents, but contain 65 percent of today's existing species. Tropical moist forests cover 7 percent of the continents, and are the habitats of around 50 percent of all species, especially the ones not yet indentified but only estimated. Most of them are insects, with some 100 000 species of flowering plants (Table 23).

From the point of view of global distribution, the role of the one species Homo sapiens is self-evident. Perhaps less well-known is that the family of grasses, within the only 25 million-year old plant family, with about 10 thousand species is not only the largest family of plants, but also the most widely distributed. Half the surface of the Earth is covered by grasses. The emergence of grasses made feasible the emergence of the unguals—the hoofed animals. Today grasses and the grass-consuming unguals are by far the most important plants and animals, in terms of human welfare.

Human activity is probably the most dangerous unique phenomenon in the long history of mass extinctions in the whole of evolution. This point requires no commentary.

Of course, throughout the long period of biogenic evolution the extermination of species has proceeded without interruption, but the present extermination rate is alarming. Parallel to this, human activity seems to have the possibility of also being

Table 23. Number of species existing at the present time

Class	Identified species	Estimated species
Animals, total	1 340 000	4 450 000
Mammals	4170	4300
Birds	8715	9000
Reptiles	5115	6000
Amphibians	3125	3500
Fishes	21 000	23 000
Invertebrates		
insects	1 000 000	3 000 000
others	300 000	1 400 000
Plants, total	400 000	480 000
Vascular plants	250 000	280 000
Nonvascular plants[a]	150 000	200 000
Fungi[b]	2000	3000
Microorganisms[b]	200 000	300 000
Present living species	1 942 000	5 230 000

[a] Nonflowering plants, mosses, lichens
[b] Very roughly

a positive element in the further evolution of the terrestrial biosphere. Man can assist in the forestation of deserts, even if the deserts are the product of his own destructive activity in the past and at present. Man has already generated, by classical breeding techniques, new cultivated species of plants and animals. There exists a small, but not negligible probability that gene-technology will, among other achievement result in the formation of new species of plants and animals, which will enrich the realm of living things on this planet. For the first time Nature influences its own evolution not solely by random activity, but by more or less conscious intelligent action. And this need not only have a negative impact on the richness of life on this planet.

Biogenic Limits of Food Production?

Ten thousand years ago, Man the hunter/gatherer consumed approximately 100 different edible plants, roots, and fruits, and about 50 different types of flesh, mostly having little fat. This was sufficient to give him all he needed in the way of energy, structural material, and essential trace elements. Sufficient food of this type can be produced on this planet for a human population of only 20–25 million. However, since the last ice age, that is, 10 000 years or about 500 human generations ago a dramatic change in this situation has occurred.

Instead of the 100 different edible plants, man now mostly consumes nine vegetables or vegetable products, very rich in starch (carbohydrates), such as rice, wheat, maize, and potatoes. Instead of the 50 low-fat meats, man today consumes no more than three types of meat with high-fat content. This major change arises from the need to increase the productivity of the plants and animals consumed.

The long-term selection of domestic animal has resulted not only in increasing productivity of the animals products—meat, fat, eggs and milk—but also in a significant decrease of the brain volume of these animals of about 10–15 percent, in comparison with the wild variety. The selected plants are much more endangered by different parasites and diseases. The impact of this spontaneous selection of the genetic pool is enormous and rather negative for the further evolution of the biosphere as a whole.

The developing science of gene technology, however, has the opportunity of influencing the breeding of domesticated animals, plants and cultured micro-organisms in a positive direction and so helping the further evolution of the biosphere overall (Table 24).

The Distant Future of the Terrestrial Biosphere

It is inevitable that a discussion concerning the future existence of mankind on the Earth can only be very weakly argued and must remain inconclusive. However, in spite of this it seems that some possible or even desired directions of the future impact of mankind on the further evolution of the global biosphere can be formulated.

Some assumptions or postulates are that:

—mankind will achieve an equilibrium population of around 8 to 9 giga people. The biomass of this world population will reach roughly 0.5 Pg,
—the production of agriculture will increase, but at a far lower rate than the population increase, and will reach a value corresponding to 1/4 or so of the global primary production,

Table 24. Biogenic possibilities for human food. All values are approximate mean numbers of species at the present time and are for illustration only.

	Animals	Plants	Microorganisms and fungi[a]
Total number of species	1 500 000	350 000	100 000
Non-edible (mostly insects)	1 000 000	270 000	95 500
Edible species	500 000	80 000	500
Noncultivated species	499 900	79 500	
Cultivated species	100	100	50
Economically significant species	50	40	
Very significant types	10[b]	12[c]	

[a] only order of magnitude
[b] including: cattle, pigs, sheep, horses, poultry, fish
[c] including: rice, wheat, corn, potatoes, cassava, sugar, oil seeds, soyabeans, bananas, apples.

—the use of other food sources, such as microorganisms, as protein source, and the use of industrial chemical synthesis of fats, amino acids and vitamins etc., will expand,

—the energetical basis for additional food production will rely on the direct use of solar energy, on the chemical energy of gas and oil products, and on the direct use of electricity from other sources, e.g. solar, nuclear etc.

—new species generated by means of gene technology will have higher efficiency of conversion of energy into protein, fat, vitamins, etc.

All these developments could enable the damaged biosphere to be restored and could even lead to its significant quantitative and, especially, qualitative improvement.

Man and the Flow of Energy and Matter: Past, Present, Future

Man and the Biosphere: Past, Present, Future

Global Human Population

Five hundred thousand years ago the 'fire revolution' began to illuminate the path of Homo habilis to new possibilities for survival on this planet. The size of the global population of hominids at this time is very difficult to assess but was probably between 10 and 100 thousand.

Fifty thousand years ago Homo sapiens set out on his way to becoming master of the world. He settled down in Europe, then in Asia. Twenty thousand years later he reached North America and Australia. Ten thousand years after the Pleistocene epoch, all regions of the world, with the exception of the extreme polar regions, the highest mountains and the driest deserts, were occupied by human hunters and gatherers. The global population was then of the order of magnitude of one million. At this time, the sea level all over the world rose as the glaciers of the last ice age, which ended 10 thousand years ago, melted, converting many lower coastal and river systems into tidewater estuaries. These coastal estuary systems were very rich in marine food resources. Here the first human settlements began.

Other settlements arose in the Near East in areas where wild wheat and barley grew naturally and yielded annually approximately 500 kg of grain from different grasses per hectare; that is, 0.05 kg/m^2. This allowed a family of four persons to harvest in three weeks a year's food supply of grain, and resulted in the 'agriculture revolution.' At this point the increase of the world's population took off. By 8500 B.C. permanent villages existed in the Middle East with well-developed storage facilities. Between 7000 and 3000 B.C. The population in the Middle East increased sixtyfold. This is equivalent to a doubling time of 650 years, or an annual increase of 0.11 percent. During the later centuries the annual rate of increase of the global population remained at the same level of about 0.1 to 0.2 percent. A dramatic step occurred in the 18th century, during the 'first industrial revolution', as the annual global population growth rate increased significantly. In the middle of the 'second industrial revolution' (the middle of the 20th century) the annual rate reached some 2 percent and probably achieved its maximum level. In the coming period of the 'third industrial revolution' (the 'information revolution') a reduction of the annual rate seems probable. The value in the mid-eighties had already decreased to around 1.7 percent per year [67, 68] (Fig. 24).

The biosphere at the present time has a dry mass of $1841 \cdot 10^{12}$ kg. The dry mass of all animals, marine and continental, equals $50 \cdot 10^{12}$ kg, which corresponds to $115 \cdot 10^{12}$ kg of 'living mass'. Present mankind, with roughly $5.2 \cdot 10^9$ people (5.2 giga), has a mass of about $0.2 \cdot 10^{12}$ kg. corresponding to 0.17 w% of the animal mass. This is probably the first time in the long history of the evolution of the terrestrial biosphere that one species alone comprises such an extremely large proportion of the total animal realm. This alone shows the uniqueness of the current global situation.

However, the most important present feature is not the absolute value, but the fact that the annual rate has been very high—roughly 2 percent. Recently, the annual rate has been decreasing, and at present equals 1.6 percent annually. The corresponding doubling time is thus 44 years. If the annual rate of increase of the world population equals 1.5 percent, after around 50 years the population will reach roughly 10 giga, and after the next 50 years 20 giga. There is no question that the most important problem facing mankind is the control of the population explosion (Fig. 25).

Mankind in the Distant Future

There is no strong basis for making a long-term prediction of the future expansion of the global population. In spite of this, it is better to make a rough estimation than to have nothing at all. The best guess (a very subjective one) is that a total of 8 to 9 giga people will be reached in the next 100 years and then tend to stabilize. Of course, this number could be considered as 'optimistic' (from the point of view of a doomsday prophecy), but most investigators are not far from this prediction. In this chapter the value of 8 to 9 giga people will be arbitrarily used for the steady-state global population (Fig. 26).

How Much Surface for Mankind?

Among numerous constraints upon the quantitative growth of mankind on this planet, the size of the planet is the ultimate one. Some general considerations show that this fact is not only of local, that is terrestrial, but also of cosmic significance. Life, according to the general definition can only exist on a planet with a solid surface, a captured liquid hydrosphere and a transparent gaseous mantle. Too-small a planet, with a mass of less than 10^{21} kg, has a gravitational field which is too weak and loses its volatile components for ever. Too-large a planet, with a mass of more than 10^{27} kg, has a much stronger gravitational field and the quantity of volatile components is so large that the gaseous atmosphere loses its transparency, which deprives living organisms of photochemical reactions, such as photosynthesis. Of course, the whole problem is strongly influenced by the distance of the planet from the central star. On margin one remark: Cosmic bodies with mass of the order of magnitude of 10^{28} kg are able to start nuclear processes in

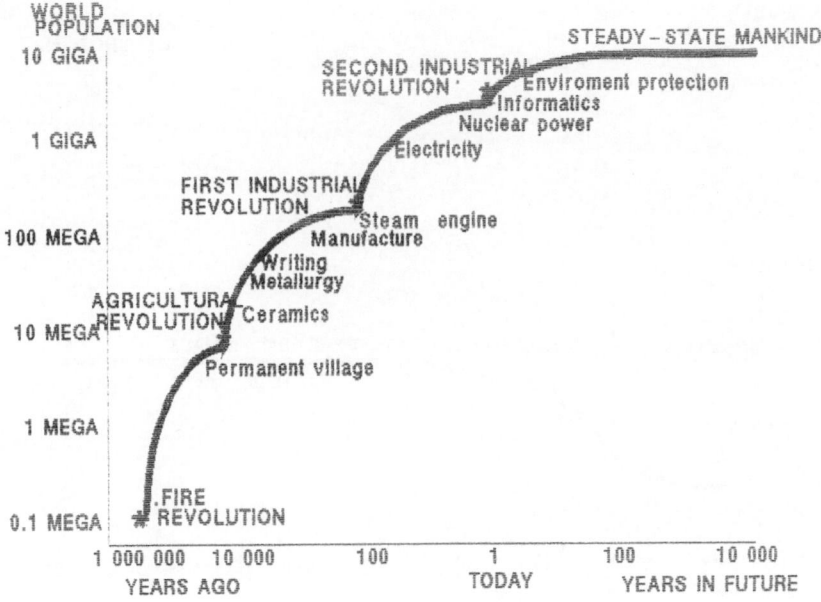

Fig. 24. Human population in the past

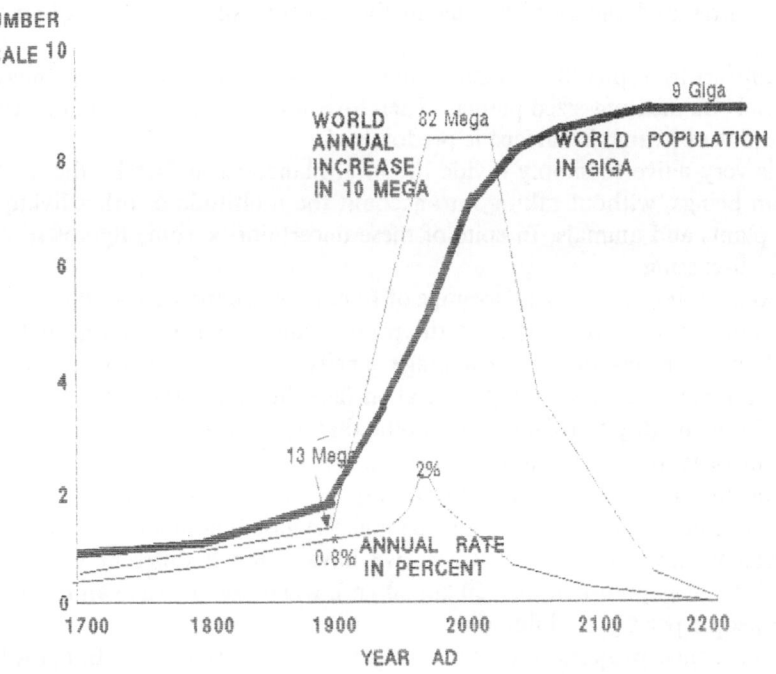

Fig. 25. Global population at present

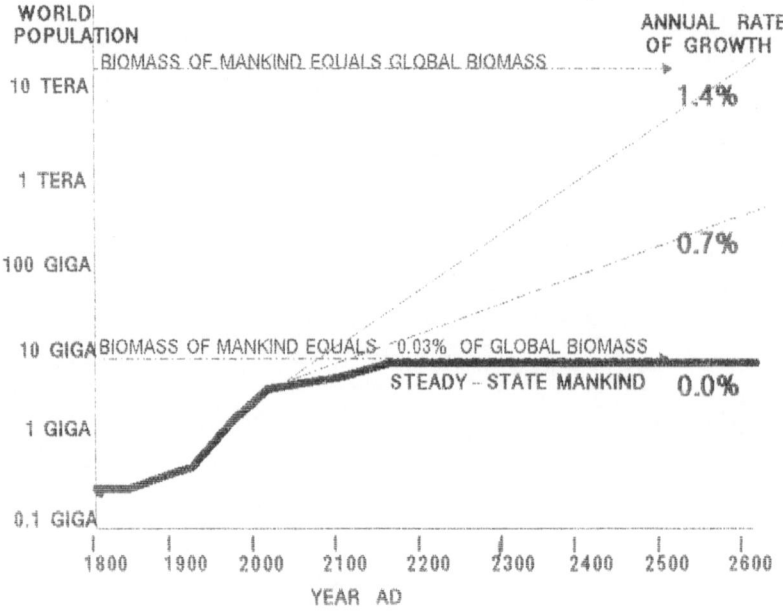

Fig. 26. Human population in the distant future

their centres and therefore belong to the category of stars and not to that of planets.

It appears to be possible to claim that living beings elsewhere in the Universe can exist only on medium-sized planets. Therefore, for each system of living beings the constraint of available surface is predominant.

It is very naive to simply divide the total planetary surface by the number of human beings, without taking into account the multitude of other living forms, both plants and animals. In spite of these uncertainties, some figures seem to be worth discussing.

About 10 000 years ago, an average of 10 million square metres were available to each inhabitant of the Earth. At the present time, each inhabitant globally has 27 000 m² of the continents: an average density of 35 people per square kilometre of continental landmass. For the total surface, including the oceans, the surface available is roughly 100 000 m² per capita; that is, 10 persons per square kilometre. The majority of the world's population lives, however, on a more restricted continental area, with an effective density of more than 50 persons per square kilometre. For an assumed 8–9 giga people, the global population density will not be dramatically different. However, we have a much different situation in a 10-million megapolis (surface: 50 km·50 km), where the density can reach around 5 000 people per square kilometre.

In some naive projections of the expansion of the human population, the bottom of the sea plays a significant role. There is, however, no serious basis for such a proposal. Even if the tremendous difficulties of living under a pressure of, say,

20 bars (equivalent to a depth of 200 metres) can be overcome, the main problems lie in other directions. The surface of the continental shelf, out to a depth of 200 m, is rather small and equals 6.5% of the total surface of the Earth. Nevertheless, although this underwater area could increase the continental area by more than one fifth, it is obvious that this cannot be the solution of the problem.

The ultimate limit to growth is the surface of the planet. All other growth limits are secondary or tertiary consequences of this primary limiting factor (Table 25).

Another Solution: Extraterrestrial Colonisation

The question of extraterrestrial colonisation belongs to the realm of science fiction, and not even the best of the genre. As a serious proposition, colonisation beyond the boundary of the Earth cannot solve the population problem, for the following simple reasons:

—Assuming that the global population equals 8 giga people and has an annual rate of increase of only 1 percent, the absolute population growth equals 80 million people per year; that is, roughly 150 people per minute.

—If mankind decided to 'export' only the annual population increase (which corresponds to a stabilisation of the terrestrial population to the above-mentioned

Table 25. How much land for a 1 million community in the distant future

Assumption: Global human population: 8 giga
Continents (excl. Antarctic): 133 million km^2
Global average population density: $60/km^2$

Region	Components	Surface km^2	Square Side, km	Population density per km^2
Dense Agglomeration		10 000	100	100
	Urban agglomeration Industrial etc. zone			
	Power station, 8 GW	2 500	50	90
	Rural zone	5 000	70	20
	Parks, forest	2 500	50	zero
Protected natural zones		10 000	100	[a]
	Tropical forests	2 500	50	[b]
	Tundra, savanna	2 500	50	
	Desert	2 500	50	
	Mountains	2 500	50	
Oceanic regions		30 000	173	4[c]
	Equatorial ocean	10 000	100	
	Subtropical ocean (including artificial islands	7 500	85	
	Temperate ocean	7 500	85	
	Polar ocean	5 000	70	

[a] Very limited
[b] Tourists only
[c] Assuming 1 month per year for each inhabitant

level of 8 giga people) by means of extraterrestrial colonisation, this would require the launching of a spaceship with room for, say, 1000 emigrants roughly every 7 minutes, or 215 flights daily.

—Such a spaceship would need to reach a velocity of up to 1/10 of the speed of light. Without calculating the exact amount of energy required by each spaceship, currently this is prohibitive, even to a science fiction author.

—Another absolute limit for the colonisation of other cosmic bodies was observed by F. Drake, 15 years ago. We know the number of stars in our neighbourhood to be about 30 000, inside a radius of 110 light years (Note: our Galaxy, the Milky Way, contains approximately 400 giga stars, inside a radius of 100 000 light years, neglecting the disc-like form of the Galaxy). Assume that all stars have at least one planet which is able to support the terrestrial form of life, and that each planet is capable of carrying a population of some giga humans, like the Earth.

—Then assume that the cosmic colonists also cannot control their birth rate, like their parents on the old planet. It thus follows that after some tens of years the inhabitants of the neighbouring planets would have to begin their own cosmic expansion and send their own spaceships with their own colonists. This would force the old Earth to send its own spacecraft to even more distant planets with significantly increased velocity.

—If an exact calculation is made it will be clear that, after some hundreds of years from the start of cosmic colonisation, spaceships which must be sent from the Earth must be faster than the spaceships sent by the colonists from the farthest colonies, and at the end must move with the speed of light. The absolute physical limit has been reached.

Cosmic colonisation thus comes to an end after only some hundreds of years because of physical limits. There is no positive solution for an ever-increasing population, even if its annual growth drops to only 0.1 percent. This will only push the physical limit forward in time. Limitless growth of any system even a super-intelligent one, is thus forbidden by the laws of Nature. The only solution is self-limitation. In such a case, the Earth remains a good enough place for the existence of mankind, even in the very distant future, assuming that other factors are not fatal.

Man and the Flow of Matter: Past, Present, Future

Total Flow of Matter

Statements about the Earth's limited resources have been very common in recent years (e.g. 'spaceship Earth'). Many qualified people, among them the members of the Club of Rome, have presented complex calculations concerning the future depletion of the planet's resources. However, it seems that the real situation has not been correctly understood by all, because the following arguments have not been taken into account:

—Our planet contains all stable and semi-stable (long-lived, radioactive) elements, without exception and in substantial amounts. All of them belong to the continental crust and most of them are concentrated by geochemical processes so that large-scale exploitation is possible (see Table 16).

—The gravitational field has prohibited the escape of almost all elements, with the exception of small amounts of hydrogen and helium.

—Industrial processes (except energetics based on nuclear fuels) do not destroy atoms but only change their chemical neighbourhood. The only exception is the fusion of light nuclides (H, D, Li) and the fission of the heaviest nuclides (Th, U), which destroy elements irreversibly.

—The present state of science and technology allows one to synthesize almost all compounds and minerals; the only constraint being the amount of free energy used and generally speaking, the economic price of a process.

—From the point of view of this book, a very important question is the quality and quantity of waste matter and waste energy generated during the extraction of substances from the atmosphere, hydrosphere, lithosphere and biosphere, or due to industrial synthesis.

—The ultimate constraint is not the amount of material resources, but the energy consumption required for production and recycling. Here it has been assumed that energy consumption is not limited to fossil fuels but relies mainly on renewable sources, primarily solar radiation and nuclear (fusion and fission) energy sources.

Taking all these arguments into consideration it must be said that, from the point of view of material resources, the planet Earth is able to sustain life, and even intelligent life, for a very long time, even for gigayears into the future (see also the chapter concerning the 'red giant' phase of the Sun).

The present consumption of materials (including fossil fuels) is, both per capita and globally, very high and perhaps not far from the allowable maximum. A very rough estimate of the present consumption of materials, including fossil fuels, is shown in Figure 27. The total annual material flow in the developed countries is presently around 4500 kg/capita. Excluding 1500 kg of fossil fuels, about 3000 kg/capita is due to other basic materials. Fuels are, of course, non-recyclable. The amount of non-energetic materials presently recycled and reused can be taken as zero (Table 26).

There are optimistic signs for efficient management in the future of the use of non-energetic materials, including a reasonable amount of recycling. The problem of the increase of energy used for material recycling will be discussed below. One possible, and even reasonable, solution to the problem of efficient recycling relies on the following premise: that the Earth's crust, hydrosphere and atmosphere not only contain all stable and semistable (long-lived, radioactive) elements, but that these are in such quantities that it seems to be possible to supply almost all elements (but not minerals) in amounts which more or less correspond to the conceivable consumption rate for a highly developed civilisation. In such a case, the mining of the required raw material is limited to the 'mining' of average crustal rocks. Of course, 'average rocks' do not exist, but the average composition of global 'rock mining' could be aimed in this direction. It is evident that the average waste coming from this type of civilisation corresponds to the average composition of crustal rocks. This seems to be the model for future 'perfect recycling', which

Table 26. Total global matter flow: geological, biogenic and technological cycles

Total mass kg		Massflow kg/year	
100%	Earth $5.6\cdot10^{24}$		
		10^{24}	
10%			
		10^{23}	
1%			
	Earth's crust $2.4\cdot10^{22}$		
		10^{22}	
0.1%			
	Sediments $3.2\cdot10^{21}$		
	Hydrosphere $1.37\cdot10^{21}$	10^{21}	
0.01%			
		10^{20}	
10 ppM			
	Cryosphere		
		10^{19}	
1 ppM			
	Atmosphere $5.1\cdot10^{18}$		
		10^{18}	
100 ppG			Rainfall, global
		10^{17}	
10 ppG			
		10^{16}	
1 ppG			Biogenic cycle
	Biosphere $1.8\cdot10^{15}$	10^{15}	
100 ppT			
			Crust/mantle exchange
		10^{14}	
10 ppT			Spoil from mines
	Animals		Agricultural waste
		10^{13}	
1 ppT	Technosphere		Fuel production
			Land denudation
		10^{12}	
100 ppP			Food for mankind
	Humans (5.1 giga)		Iron production
		10^{11}	
10 ppP			Ammonia production
			Aluminium production
		10^{10}	
1 ppP			

results in the ultimate solution of the global material cycling problem [69, 70].

The only difficulty, and one which is far from being harmless, is that of the waste from nuclear energy production. It is true that 'average crustal rocks', especially granites, contain amounts of uranium and thorium which allow not only the mining and decomposition of the rocks into elements, but also allow the production of energy, for example electricity, in more than adequant quantities. In addition to this, deuterium from water and lithium from the same crustal rocks could be used in thermonuclear energy systems to achieve the required level of energy consumption. However, in both technique–nuclear fission and thermonuclear fusion–significant amounts of radioactive, partially long-lived nuclides emerge. The problem of the ultimate management of high-radioactive waste is still waiting for an appropriate solution, but there exists some possibilities for achieving this in the not too distant future. This will probably be the most serious question concerning material recycling for future civilisations (Fig. 27, Table 27).

The Best is Taken Now

Our civilisation is tightly bound up with the use of metals. Our technology is first and foremost based on machines, instruments, vehicles, buildings, bridges, tubes, wires, etc. made of metals. The metallic elements must fulfil particular criteria, such as 1) desired mechanical, electrical, chemical, and thermal properties, 2) chemical stability in the atmosphere and hydrosphere, 3) ease of recovery and

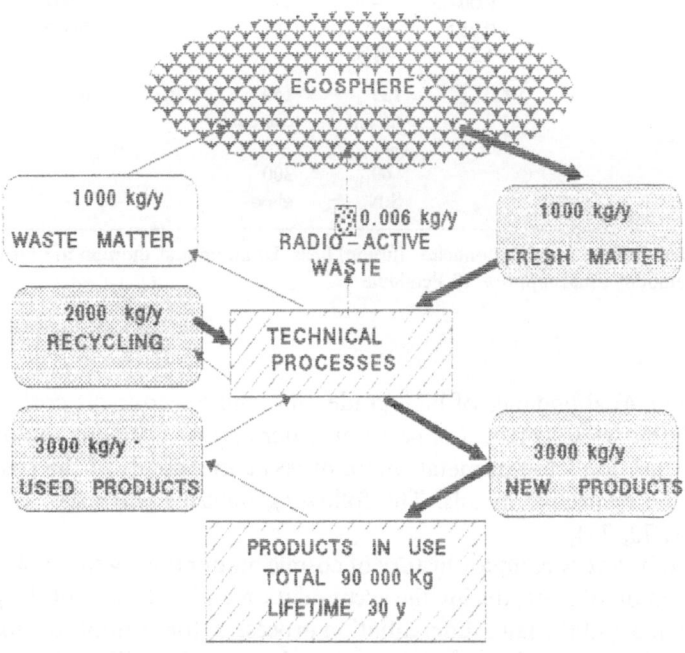

Fig. 27. Recycling of matter in the future

Table 27. Elementary content of one ton of ecosphere and its possible technological use in the distant future. Ecosphere = average of atmo-, hydro- and lithosphere.

Element	Content of one ton of eco-sphere		Potential production			Nuclear fuel	Other products
			'Metals'	Cement glass	Ferti-lizers		
	Mole	kg	kg	kg	kg	kg	kg
Oxygen	31 600	500	–	80	–	–	300 free
Silicon	9 180	257	60	300	–	–	?
			(Si_4N_3)	(SiO_2)			(SiC_xH_y)
Hydrogen	720	0.13	–	–	–	Energy carrier	
Deuterium	0.13	0.0025	–	–	–	0.0025[a]	
Aluminium	2 900	78	78	–	–	–	–
Sodium	1 050	24	–	24	–	–	–
Iron	840	47	47	–	–	–	–
Potassium	780	33	–	30	3	–	–
Magnesium	700	17	10	7	–	–	–
Calcium	575	23	–	18	5	–	–
Titanium	85	3.8	3.8	–	–	–	–
Phosphorus	34	1.1	–	–	1.1	–	–
Fluorine	31	0.6	–	–	–	–	?
Carbon	16	0.2	0.2	–	–	–	–
Nitrogen	10	0.15	0.15	–	air	–	–
Sulphur	8	0.25	–	–	0.25	–	–
Lithium		0.065	–	–	–	0.065[a]	–
Uranium		0.004	–	–	–	0.004[a]	–
Thorium		0.012	–	–	–	0.012[a]	–
Other		1.5	1.5	–	–	–	?
Total		1000	135 metals + 60 SiN	150 cement + 300 glass	10	0.1	345

[a] Deuterium and lithium are thermonuclear (fusion) fuels. Uranium and thorium are fissionable nuclear fuels. Energy amount equals approx. 10 Petajoule

manufacture, 4) abundance of high-grade ores, and 5) economic cost of preparation. With this wide spectrum of selection criteria, it is surprising to note that the present civilisation's use of metals more or less corresponds to the cosmic abundance of the respective metals. The following remarks will make the situation clearer [71, 72, 73].

In spite of the very complex history of cosmic matter during the evolution of the three generation of stars, during the accumulation and evolution of the protosolar gas-dust cloud and the later differentiation processes, for example the loss of more than 99% of the volatile components and the complex differentiation between crust and mantle, the abundance scale of most metals in the Earth's crust, even in the neighbourhood of the surface, is very similar to the cosmic abundance scale (see Table 12) [73].

From this point of view, that is, the proportion of the amount being used by civilisation to the value of the abundance level in the Earth's crust, all metals belong to one of the three following groups:

1) the 'underproportionally' exploited: e.g. silicon, titanium, aluminium

2) the 'proportionally' exploited: e.g. iron, nickel, chromium, zinc, and even the relatively rare metal, silver

3) the 'overproportionally' exploited: e.g. copper, lead, molybdenum (Fig. 28).

Quasi-Irreversible Matter Flow: Concrete, Metals

Much more severe for the future seems to be the fate of wastes in solid form, such as concrete, scrap metals, solid chemical wastes etc. To this group also belong demolished buildings, abandoned highways, etc. Our civilisation has no clear idea how to deal with this group of wastes and how to calculate the social and environmental costs of their treatment. Also, the chemical impact of solid, low soluble, but not harmless, substances over a long time period has not been adequately investigated.

Irreversible Material Flow: Fossil Fuels

Because of the physical stability of the most-used elements and because of the gravitational field of the planet, the flow of almost all matter can be reversed. The only price for reversibility is the amount of free energy needed for the appropriate

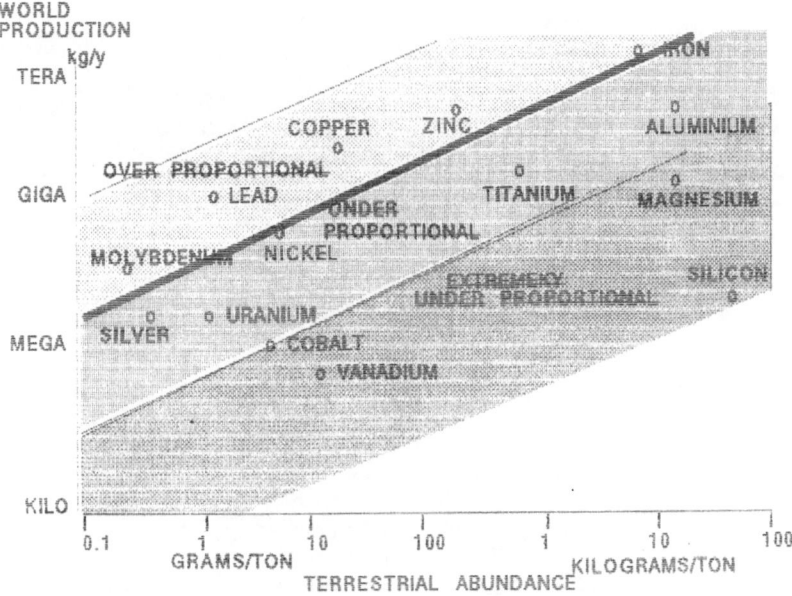

Fig. 28. Metals: cosmic abundance versus terrestrial exploitation

process. From the practical point of view, it is unreasonable to try to turn back the process of the oxidation of carbon and hydrogen. The burning of fossil fuels must be considered as irreversible. This means that, in the distant future, when the recycling of matter is one of the highest priorites, the burning of fossil fuels must be prohibited. The burning of renewable biogenic fuel, such as wood, kelp, etc, is of course reversible, due to photosynthesis; that is, due to solar radiation, which is the result of irreversible processes in the Sun's interior (see the Chapter concerning the 'nuclear burning' of hydrogen) (Fig. 29, Table 28).

Reversible Flow of Matter: Water, Food, etc.

The question of the global water cycle has already been discussed in this handbook (see Vol. 1. Part A, p. 17). This is the largest material cycle on this planet. However, in spite of this, even at the present time the elements of an acute crisis in the water supply for mankind are evident, and an acceptable solution is still waiting to be realized.

An adult drinks some 2 kg of water per day, or 0.7 cub.m per year. Adding his needs for washing, cooking, agriculture (plants and animals), and industry, as well as for cooling in power stations, the average per capita annual demand is some 1500–1800 cub.m. In the United States the annual amount is higher, at around 3000 cub.m. If the mean annual rainfall corresponds to a 0.7 m layer of water, and allowing 70% evaporation, one American needs a surface of 14 000 m² to supply his need for water. The total area available to him at present is 44 000 m². Even in

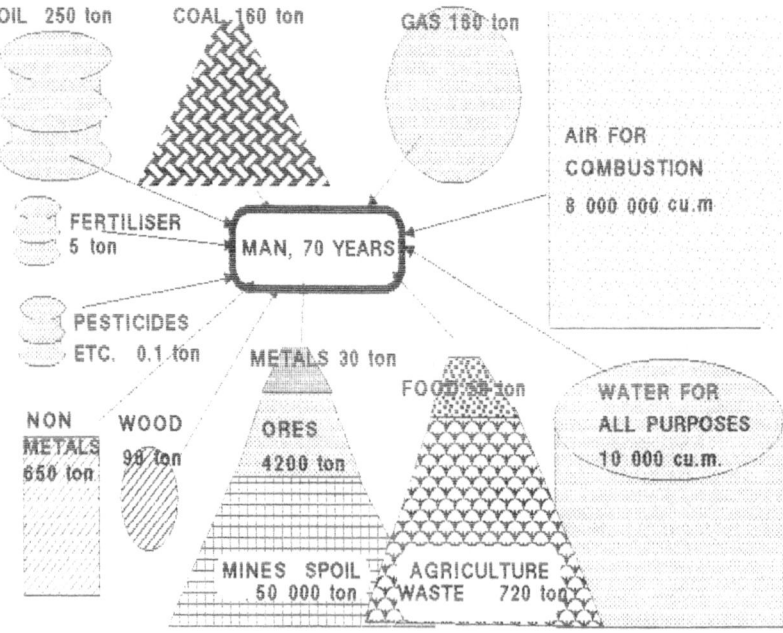

Fig. 29. How much matter does one man need in a developed country?

Table 28. Material flow per capita at present and in future. Annual material flux in kg/y. For the present, data are for developed countries; for the future, the postulated global average

Materials	Present	Far future
Fossil energy carrier (coal, oil, gas)	1500	0
Nuclear energy carrier and massless carrier	<0.1	<0.1
Air for combustion	large	0
Food, human	300	400
Organic materials (wood, paper, fibres)	300	150
Metals	200	200
Anorganic materials (cement, fertiliser)	350	350
Sand, etc.	2000	1000
Total materials	4500	2100
Recycling	Insignificant	Significant

the next decade, when, by the year 2000, the world population will reach more than 6 giga people and the average annual use will increase to around 1000 cub.m the total use of water will increase to approximately 1 percent of the total rainfall over the continents and about 3 percent of the total flow of water in all the world's rivers during one year.

Natural human and animal food, being a product of photosynthesis, is a part of the full, reversible flow of matter. However, some limitations to reversibility are present even here, such as the use by agriculture of some fertilizers which are the products of mining of, for example, phosphates or potassium salts which have been accumulated over a long period and now are being dissipated and, at the end, dissolved in the oceans. This kind of irreversibility will in the future, determine the character of terrestrial material recycling.

Another cause of irreversibility in the field of food production for humans and animals is the growing proportion of biosynthetic and chemosynthetic nutrients (see Chap. below), which are coupled with specific wastes, which may influence material recycling.

Matter Recycling and Energy Use

The ultimate reason for the irreversibility of matter is the amount of free energy used for the regeneration of the previous physical and/or chemical state of matter. The concept of recycling relies on the assumption that the environmental impact of free energy used and the associated material and thermal waste is significantly lower than the impact of the material waste which is the object of the recycling process. Therefore, knowledge of the energetical price of the recycled material is the most important factor in the process. (Table 29).

Table 29. How much energy for production and recycling of materials?

Production from raw materials MJ/kg	Metals	Chemicals	Naturals	Recycling from waste MJ/kg	Recycling from waste (material)
	Metals				
	Uranium				
500	Titanium			500	
	Magnesium				
	Aluminium				
	Nickel				
160	Special steel			160	
	Copper	*Chemicals*			
	Silicon	Nylon			
	Steel				
	Zinc				
50		Ammonia		50	
	Lead				
					Magnesium
		Paper			
		Glass			Copper
16				16	
					Iron
		Plastics			
		Soda			
		Cement			
5				5	
			Naturals		
			Clays		
1.6				1.6	
			Sand		
			Gravel		
0.5				0.5	

Global Flow of Materials in the Future

Based on the more or less arbitrarily estimated values given above for the material flow per capita and the size of world population, it is possible to make some estimates concerning the distant future. Without going into details, among other reasons because of lack of precise information, the following prediction can be put forward.

Within the next two or four decades the global material flow will increase, due to increasing use of per capita in the developing countries and due to increasing global population. Later, the rate of annual increase of the material flow will steadily diminish. Within one or two centuries the global material flow will slowly begin to decrease and reach a level which will probably not differ much from the present. The quality of materials and the ratio of recycling will, of course, be the principal differences.

Relying on the assumption concerning the effective use of materials, the long life of used products and the rational recycling of matter, it would appear to be permissible to predict that the material flux in the distant future will not be much larger than at present. Of course, the composition of the material flux must be rather different, and nearer to the average composition of the continental crust.

Food for Everyone

How much Food does a Man Need?

A human, the typical heterotrophe, needs a continuous supply of, 1) material carriers of free energy for maintaining his internal order and for carrying out work, and 2) material for maintaining the structure of his body.

Let us consider first of all the free energy flux in the food required by an average man—about 110 watt. This corresponds to 110 joules per second, or approximately 10 MJ per 24 hours, that is around 2300 kcal/day. This flow of free energy is used for the following purposes:

—brain activity	15–20 W
—gastrointestinal tract	15–20 W
—heart activity	5 W
—kidney activity	8 W
—other metabolism	15 W
—muscles:	
body movement	30 W
long period work to outside	20 W
Man, adolescent, average constant flux	110 W

Man, adolescent	
—writing	10–20 W
—working	50–80 W
—walking	100 W
—working hard	300–400 W
—climbing	800 W
—sport exercise	1000 W for less 1 minute
—sport exercise	600 W for 1 hour

All these values correspond to average activity over a long period of time. Over short periods man can increase his activity to much higher levels; for example, during a march at 6 km/hour and with 20 kg load, the energy flux rises to some 600 W, which is 5 times greater than the mean energy flux of 110–120 W. Under these conditions the energy used for walking is only 200 W, leaving the remaining 400 W for more intensive metabolic processes. Such intensive work is possible for only a few hours per day and with higher than average food intake.

The second point to be considered is the inflow of structural material which is needed to sustain internal order, to remove destroyed elements of the internal structure and to replace material waste. Man needs a more or less continuous inflow of the following materials in well-defined states:

—8 amino acids, the so-called 'essential amino acids': leucine, isoleucine, lysine, valine, methionine, phenylalanine, threonine, and, for babies, also histidine. The remaining 12 amino acids needed for the body's structure can be synthesised in the human body itself,

—3 essential fatty acids: arachidonic, linoleic, and linolenic,

—14 vitamins: B-1, B-2, B-12, C, A, D, E, K, niacine, and others,

—17 elements: Na, K, Mg, Cl, P, S, Fe, F, Co, Cu, Mn, Mo, Se, Zn, Ni, Sn, and Al. This does not include the main components: H, O, C and N.

—water as an internal medium,

—air as the carrier of oxygen (Table 30).

The problem of the type of energy sources required by living beings has been discussed previously in Chapter 3. We have mentioned that Homo sapiens has changed himself from a typical 'heterotrophe' to a 'technotrophe'. Let us consider the problem of the future production of human food by means of all conceivable technologies: traditional agriculture, that is, biogenic production, improved agriculture (by means of gene technology) and out-of-soil production, by bio-synthesis (on the basis of unicellular organisms) and chemosynthesis (abiogenic synthesis).

The metabolic rate of mankind at present (5.1 giga people), assuming roughly 80 Watt per capita and including children, corresponds to 0.4 TW (terawatt = 10^{12}). Assuming that agricultural production is about 3–4 times larger then the

Table 30. Daily inflow of food components

Carriers of free energy	Grams per day
Non-essential fats	150
Non-essential proteins	70
Carbohydrates	270
Subtotal—energy carriers	490
Oxygen	860
(from 3500 g/d of air)	
Internal medium	
Water	1500
Structural material	
Essential amino acids	6.5
Essential fatty acids	6.0
Vitamins and protovitamins	
14 vitamins	0.1
Minerals	
approx. 20 elements	10
Subtotal essential materials	23
Total food per day (Fresh form)	2900
(dry biomass)	500
–corresponding to an energy of	10 MJ/day

direct intake of food, the directly consumed 'biogenic flow of free energy' reaches a value of 1.3 TW, about 1 percent of the global primary production of the terrestrial biosphere [63, 74].

The total flow of biogenic wastes resulting from agriculture, the food industry, housekeeping, and last but not least, humans and livestock, cannot be fully recycled for agricultural use and influence in high degree the soil, waters and air.

The total amount of agricultural products consumed by humans must be corrected by the amount of these products consumed by animals which are a source of human food (directly, such as meat, or indirectly as milk and eggs). The quantity of this kind of food equals roughly two and a half times the amount directly consumed by humans. In developed countries this ratio reaches a value of 6 to 8. The losses during the preparation of human food, from the field to the stomach, is probably more than one-third of the food consumed.

Taking all these nutrient flows for the global population, the total flux of energy in the form of biogenic products comes to a value of some 5 GW, which corresponds to about 5–6 percent of the total annual production of the global biosphere. Some calculations, which also take into account other indirect influences, mostly negative, and direct impacts, such as the production of nonfood agricultural products, losses due to urbanisation, etc., suggest that mankind uses more than 15 to 20 per cent of the global biogenic annual production.

The further development of food production for humans will decide the state of the environment, not only on the local, but also on the global, scale. To give an idea of the magnitude of the problem, 9 giga humans produce roughly 3 Pg of excreta, which must be processed with minimal negative environmental impact. The processing of the excreta of domesticated animals is another problem (Fig. 30).

Food Production and the Use of Technological Energy

Solar energy is not the only source of free energy in modern agriculture. The inflow of technological energy is also present in direct form—fuel for vehicles and other engines, oil for heating and drying, electricity for machines and for cooling—and in indirect form—energy content of fertilizers, plant and animal protection substances, and energy used for the production of machinery and buildings, etc.

Continental Food Production: Agriculture

Agriculture uses at present about 18 percent of the continental land surface (excluding Antarctica). This surface has been taken in the past mostly from grassland and forests (see Fig. 31). This cultivated surface, assuming a world population of 5 giga, corresponds to approximately 4600 m² per capita; that is, 0.46 hectare/capita. In the year 2000, with around 6.1 giga people, 15.4 Mm² of cropland will be in use. This corresponds to around 2500 m² of cropland per capita. The growing population will tend to reach a steady-state level of about 8 to 9 giga people. The increasing need for food, especially the increase of meat production, and non-food agricultural products calls for more cultivated areas. The danger for the natural environment cannot be ignored.

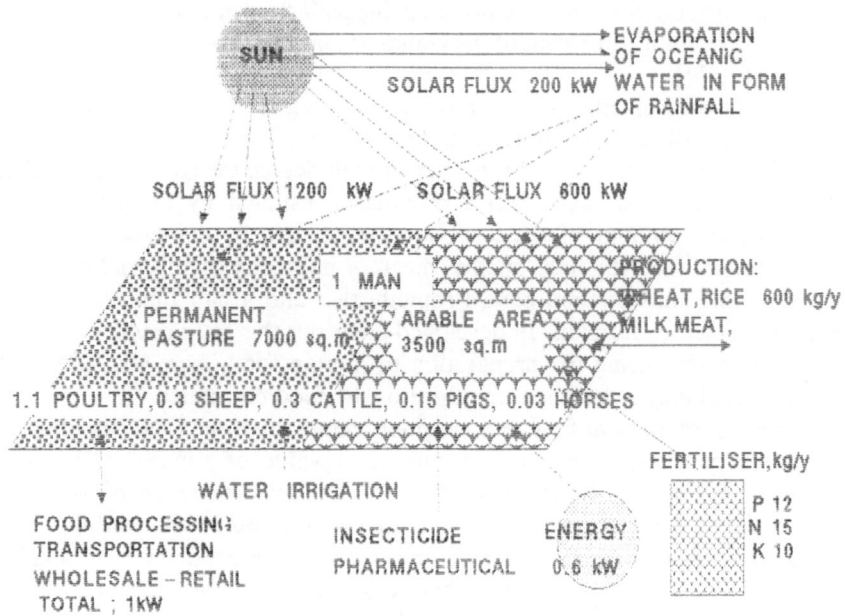

Fig. 30. Flow of energy and matter in food production

There are some 'soft' methods for increasing the extent of agricultural areas. One of them is to utilise the 10 Mm² in Africa excluded from cattle production because of the tse-tse fly. In this area the production of meat is some 70 times lower than that in Europe. In the Soviet Union the 'tcharnozem' are intensively agriculturally exploited regions which are only 8.6% of the total surface of the country. The 'tcharnozem' compose 60% of the total agricultural land in the country and generate 80% of the total agricultural production. In China the agricultural areas cover only 10% of the total territory, with roughly half of them being irrigated (Fig. 31).

Industrial Synthetic Human Food

Biogenic production of human food, both in agriculture and marine culture, will in the future occur step by step with, and complemented by, the purely industrial synthesis of food components, such as individual amino acids, essential and non-essential, special carbohydrates, essential fats, provitamins, vitamins, anorganic components, as well as synthetic meat and even synthetic caviar.

Even at present the number of synthetic human food components is growing and increasing in significance on a global scale. Industrial production methods will cover a wide spectrum, beginning from microorganismic synthesis (natural and genetically influenced), through enzymatic to purely chemically synthetic processes.

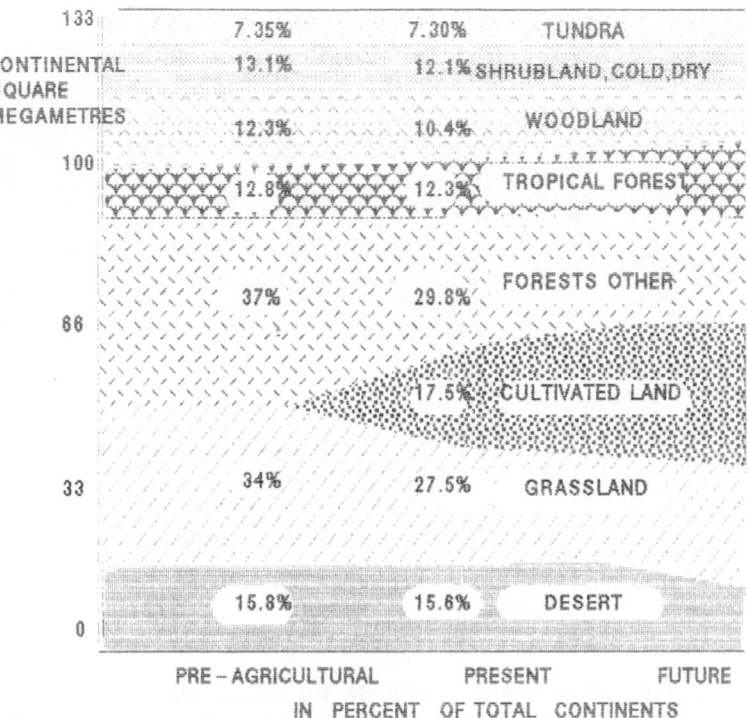

Fig. 31. Land use and cover; past, present, future

Global Climate: Past, Present, Future

Global Climate Machine

The total solar energy flux arriving on Earth equals 175 340 TW. This corresponds to a solar constant equal to 1367 W/m², a value which has been measured with an accuracy of approximately 0.2%, corresponding to ±350 TW. Recent measurements suggest that the Sun is 0.1% more luminous at the peak of the 11-year solar cycle, which corresponds on increase of about 175 TW. It is possible, but far from well established, that during the Little Ice Age (300 years ago) the diameter of the Sun was larger, and therefore its outer temperature was lower.

Assuming an albedo of the Earth equalling 0.310 (see Table 32), the effective solar energy flux absorbed by the Earth is only 120 980 TW. In addition, the value of the terrestrial albedo is not exactly known, which results in further inaccuracy in the amount of energy absorbed by the Earth. For the sake of illustration, the present total flow of energy due to human activity equals around 10.5 TW (see Fig. 32).

The global climate is one of the most complex terrestrial natural phenomena and at the same time, one of the most significant parameters influencing our future. Recently, rather significant features of the global climate have begun to be better

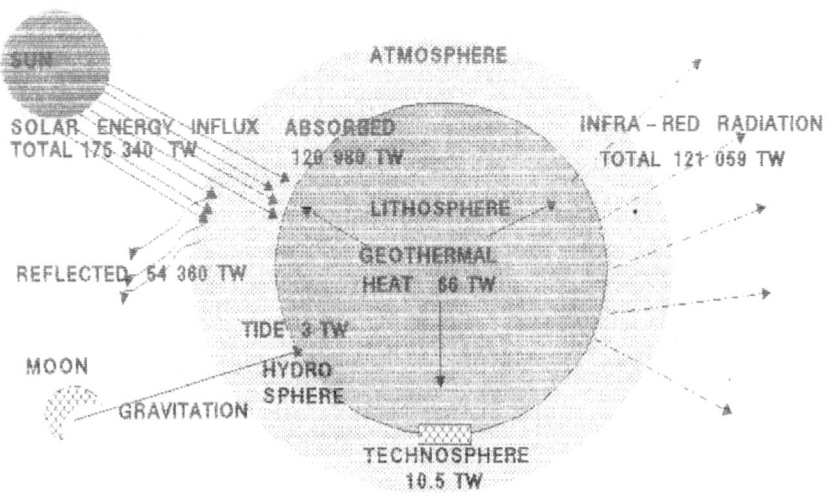

Fig. 32. Terrestrial energy balance at present

understood. Some of these phenomena are: the disruption in the ocean and wind systems in the tropical Pacific, known as 'El Nino'; the role of global oceanic currents; and the stratospheric winds influenced by the solar cycles [81, 82, 83]. Also, the coevolution of the global climate and the biosphere has been spectacular, but somewhat overestimated. This is known as the 'Gaia' hypothesis (Table 31, 32).

Sources of Free Energy on the Earth

Needs for Technological Energy: Past, Present, Future

Life is a highly ordered open system needing a more or less stable inflow of free energy and outflow of bonded, waste energy. Intelligent life uses not only the natural sources of free energy—solar radiation—but also the artificial—the technical sources of energy. The flow of energy on the Earth is shown in Table 33 [88].

The energy flowing on the Earth at the present time has different origins. All the four elementary forces—strong (nuclear), electromagnetic, weak and gravitational—are involved as producers and/or as carriers of free energy (Table 34).

The Sun, as an average 'main sequence' star, has produced approximately 10^{26} watts since 5 gigayears. Today the energy flux equals $3.72 \cdot 10^{26}$ W and will probably not change much during the next 4–5 gigayears; that is, to the moment of catastrophic transformation into a 'red giant star'. The Earth receives only a $4.7 \cdot 10^{-10}$ part of the total solar radiation flux; that is, 175 000 TW. From this flux only 121 000 TW are absorbed by the Earth, the rest being reflected back to interplanetary space. The solar energy flux on the Earth is used, among others, in

Table 31. Impact on climate: Milankovitch hypothesis and others. y = year; kyr = kiloyear; Myr = Megayear; Gyr = Gigayear

Cause	Parameter	Details	Duration	Consequences
Sun	As a star	Solar spots	11 y	Colder?
		Solar flares	?	?
		Star evolution[a]	Gyr	Changes in albedo and photosynthesis
	As a star in Galaxy	Oscillation[b]	28 Myr	Possible impact of comets
Sun-earth mutual position	"Milankovitch Hypothesis" Eccentricity of orbit obliqueness of axis Precession of perihelion		100 kyr 41 kyr 23 kyr	Significant impact
Earth as planet	Day/night rotation increases from present 24 hours to much longer period of 30 hours		Gyr	In distant future
	Atmosphere	Greenhouse effect (carbon dioxide, dust, ozone) Volcanic activity	tens of years years	Significant almost irreversible
	Biosphere	Vegetation	kyr	Albedo change
	Hydrosphere	Ice sheet	kyr	Albedo, Sea level
	Lithosphere	Continental drift (cycle hypothesis) Mountains growth Oceanic circulation changes	450 Myr	Extremely important in distant future
	Technosphere	Pollutants in: sea water, lithosphere, biosphere, air	100 y	Extremely important at present

[a] Solar evolution results in an increase of concentration of heavy metals, resulting in increasing ultraviolet radiation
[b] Oscillation of the Sun perpendicular to galactic plane resulting in crossing gas-dust clouds

the following ways:

—direct heating of the atmosphere   55 000 TW
—energy flow in the global evaporation of water   44 000 TW
—total gross energy flow converted into chemical energy
   in the biosphere (gross)   22 000 TW

166   M. Taube

Table 32. Terrestrial albedo

Surface type	Albedo A	Portion of surface f	Reflection R = A·f
Water	0.04	0.353	0.014
Clouds	0.49	0.47	0.23
Vegetation	0.13	0.13	0.016
Rocks	0.15	0.021	0.03
Ice, snow	0.70	0.03	0.02
Cities, etc	0.15	0.002	0.0003
Total		1.000	0.310

Albedo changes as a result of direct human activity

Changes caused by human activity	Albedo (A) Before	After	Portion of global surface (f)
Savannah → Desert	0.15	0.35	0.018
Savannah → Agriculture	0.15	0.20	0.14
Forest → Grass	0.12	0.15	0.016
Grass → City			0.022
Total	0.16	0.19	0.17

Total reflectivity change: $R = A \cdot f = 0.17 \cdot (0.19 - 0.16) = 0.02$

—total energy flow in the technological activity of mankind,
  of solar and non-solar origin   10.5 TW
—energy flow in the total metabolism of mankind   0.5 TW

Beginning with the fire revolution, probably 500 thousand years ago, the needs of Homo erectus for his technical activity (cooking, illumination, warming) increased more or less continuously. His own, that is his bodily, ability for doing work can be estimated at around 30 W over a period of many hours, with the possibility of increasing this power by a factor of 5 or even 10 times over a short period. The energy flux emanating from fire is here assumed to be roughly 300 W.

How Much Energy Does a Man Need?

Indirectly, what man ultimately needs is order. He needs order in manifold forms and in manifold ways. What is meant here by 'order'? In the first place, man needs order in his own structure, in his body. Only the more or less continuous intake of free energy prevents the spontaneous decay of internal order, the highly complex living and thinking system. For this purpose man needs a constant intake of free energy carriers—in the form of food—of about 10 MJ per day: that is, about 2400 kcal/day, equivalent to:

$$\frac{2400 \text{ kcal}}{\text{day}} = \frac{10 \text{ MJ}}{\text{day}} = \frac{10^6 \text{ J}}{86\,400 \text{ s day}} = 116 \text{ W}$$

Table 33. Energy flow: a scale

Watts		Watts		Watts	
10^{28}		10^{16}		10^4	
	Sun as star				
			Photosynthesis		Man
10^{26}		10^{14}	global	10^2	metabolism
					Bulb
10^{24}		10^{12}		1	
			Moonlight		
			Mankind's		
			metabolism		Pocket
			Volcanoes		calculator
10^{22}		10^{10}		10^{-2}	
			Nuclear power		
			station		
			Stellar light		
10^{20}		10^8		10^{-4}	
10^{18}		10^6		10^{-6}	
	Fusion bomb				
	Solar radiation		Automobile		
	Global climate				Micro-
					organism
10^{16}		10^4		10^{-8}	

Table 34. Elementary forces as energy source the earth at present

	Elementary forces			
	Strong	Elec-magnetic	Weak	Gravitation
Stored energy		Big Bang Deuterium (in seawater) Giga terawatt-years		
		Supernova Thorium + Uranium (in lithosphere): Giga terawatt-years		
Continuous energy flux		Earth's radioactivity Geothermal heat: 66 TW		
		Solar radiation Photons, resulting from hydrogen fusion in centre of Sun: 121 000 TW		
				Moon Tidal energy: 3TW

This amount of energy for internal metabolic needs corresponds to a naked man at the level of about half a million years ago. (Table 35, Fig. 33).

The value of 300 W (that is, 300 J/s) per capita should be understood in the following way. If an energy system generates an average flux of 300 watt over a very long period, this means that during a year it will generate:

$$(300 \text{ J/s}) \cdot (3.15 \cdot 10^7 \text{ s/y}) = 9.45 \cdot 10^9 \text{ J/y} = 9.45 \text{ GJ/y}$$
$$1 \text{ Watt-year} = (1 \text{ Watt}) \cdot (3.15 \cdot 10 \text{ s/y}) = 31.5 \text{ MJ/y}$$

In general wood, in most cases, was gathered in the direct neighbourhood of dwellings. During the long period of some half-million years, no change in the techniques of energy production occurred. Only 5 thousand years ago did new energy sources come on the scene: first the power of draught animals, later the power of falling water, and then the energy of the wind. The total amount of energy flux at this time is estimated to be at the level of one kilowatt, in the most developed countries. The discovery of coal as a carrier of energy in the form of high-temperature heat was of such significance that it triggered a new era—the industrial revolution. The energy flux in the most developed countries of this time then increased to roughly 2–3 kW per capita.

At the present time, the energy flux used by man in developed societies varies between 4 to 11 kW per capita (more exactly, 4 to 11 kilowatt-year per year). For further consideration, a value of 7 kW per capita has been taken as typical (see Table 35). It is very difficult, and open to misinterpretation, to try to describe the

Table 35. How much energy does a man need in a developed society?

Main aims	Detailed aims	Watts/Capita
Food	Agriculture, food industry, distribution, cooking, cooling	800
Environmental heat	Space heating, air conditioning	1000
Housing	Construction of houses, hospitals, schools, etc. and maintenance	500
Transport	Construction of vehicles and roads, fuel for transportation, maintenance	300
Culture	Culture, science, school system, free-time activity, information media, religion	700
Social	Social help, state organisation, fire protection, health protection, police, military	700
Natural environment	Protection of natural environment, management and recycling of waste, purification of air, water, natural parks, etc.	1500
Subtotal	Energy for direct use	5700
Energy generation	Construction of mines, pipelines, refineries, power stations, transportation of fuels, electricity, processing of fuel, search for new energy sources, and conservation of energy	1300
Total	All energy needs	7000

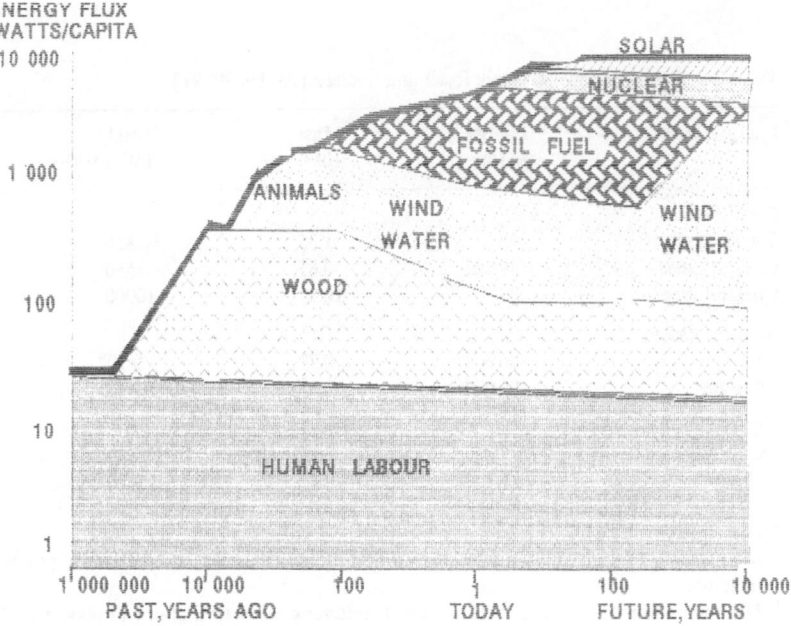

Fig. 33. Energy for man in the past

real energy needs of one average man, even if we limit ourselves to the situation in a modern developed civilisation. However, in spite of this, for the sake of illustration some figures are given in Table 35.

The energy flow has varied very strongly during the evolution of civilisation. On one hand, the kind of utilized energy has changed, and we have moved from its use for cooking and space heating to its use for technology, transport, and, last but not least, to the production of energy itself.

Non-Renewable Resources for Energy: Past, Present, Future

The first jump from the use of the renewable energy carrier, wood, to the non-renewable energy carrier, coal, was possible on this planet because of the existence in the Earth's history of the Carboniferous Period and other short periods during which copious amounts of coal were deposited. Are there cosmic civilisations which do not have access to such stored energy carriers? How then would be the evolution of energetics? (Table 36, Fig. 34).

The present situation in the area of energy perspectives is commonly described as very critical. However, from the point of view of the amount of stored energy—fossil and nuclear, fission and fusion, the direct use of solar energy and its indirect

Table 36. Stored energy on earth: fossil and nuclear [88, 89, 90, 91]

Energy carrier	Energy MJ/kg	Mass 10^{15}	Total TWyear (h)
Fossil fuel			
Oil, all types	45	1.35	850
Gas, all types	39/cub.m	0.87	550
Coal, all types	29	10.8	10000
Nuclear, fission			
Uranium	86 000[a]	0.7[b]	$1.9 \cdot 10^9$
Thorium	86 000[c]	2.8[d]	$7.9 \cdot 10^9$
Nuclear, fusion			
Deuterium	275 000	0.2[e]	$1.9 \cdot 10^9$
Lithium	275 000[f]	0.2[g]	$1.9 \cdot 10^9$

Notes:

[a] After transformation of Uranium-238 into Plutonium-239 in fast breeder fission power reactors

[b] Uranium abundance in the crust of 3 ppM, assuming extraction of one thousandth of the Uranium

[c] After transformation of Thorium-232 into Uranium-233 in thermal fission power reactors

[d] Thorium abundance in the crust of 12 ppM, assuming extraction of one thousandth of the Thorium

[e] Deuterium abundance in water of 16 ppM, assuming extraction of one hundredth of the Deuterium

[f] After transformation of Lithium-6 into Tritium in thermonuclear (fusion) power reactors

[g] Lithium abundance in the crust of 30 ppM, assuming extraction of one ten-thousandth of the Lithium

[h] TWyear = Terawatt over one year $= (10^{12}\ W) \cdot (3.15 \cdot 10^7\ s/y) = 3.15 \cdot 10^{19}$ Joule $= 31.5$ EJ

Coal world resources and reserves (in $Pg = 10^{12}\ kg = $ Gigaton)

	Bituminous	Sub-bituminous	Brown, lignite
Proven[a] amount	920	260	340
Proven recoverable reserves	515	170	265
Accessible in coalfields	360	25	100
Subtotal	1795	455	705

Total resources $= 2956\ (= 2.956 \cdot 10^{15}\ kg)$

[a] Proven reserves are quantities which geological and engineering information indicate with reasonable certainty can be recovered in the future from known reservoirs under existing economic and operating conditions.

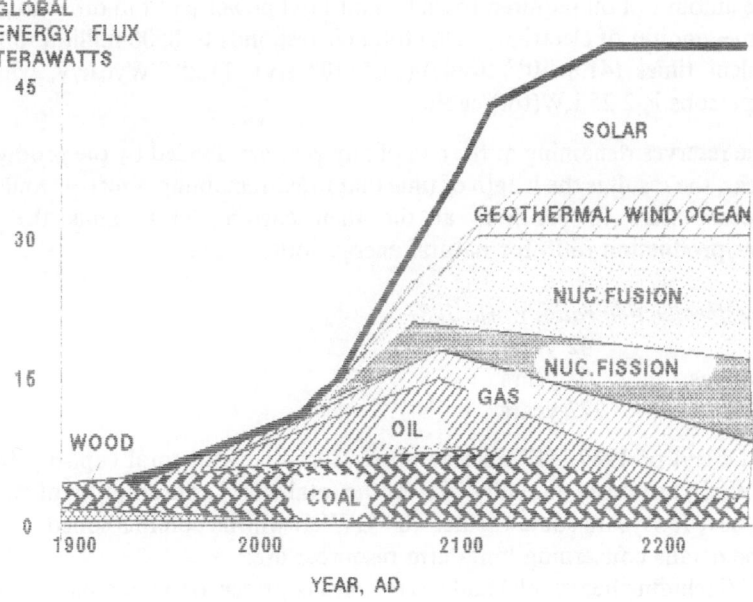

Fig. 34. Global primary energy flow at present and in future

use from wind, heat of the oceans, etc.—and the practical and economical usage the following relationship should be noted:

> 1 million tonnes of oil equals approximately:
> $41.8 \cdot 10^{15}$ Joules of thermal energy
> or 12 Terawatt-hours of thermal energy
> or 1.5 million tonnes of coal
> or 3 million tonnes of lignite
> or 1.11 giga-cubic-metres of natural gas

Some recent data concerning global energy production and consumption are worth mentioning. The consumption of world-wide primary energy 1987 was:

	In million tonnes of oil-equivalent	Percent of total
Oil	2941	29.3
Natural gas	1556	18.3
Coal	2386	28.1
Hydroelectric	524*	6.2
Nuclear	404*	4.8
Subtotal	7811	91.9
Wood, dung, etc.	690*	8.1
Total	8500	100.0

*) The amount of oil required to fuel an oil-fired power plant in order to generate the same amount of electricity. This total corresponds to 8500 million tonnes oil-equivalent times $(41.8 \cdot 10^{15}$ Joule$)/(31.5 \cdot 10^6$ s/y$)$. 11.28 TWyear/y, which, for 5 gigapersons is 2.25 kW(tot)/capita.

If the reserves remaining at the end of any year are divided by the production in that year, the result is the length of time that those remaining reserves would last, if production were to continue at the then current level. Thus the present reserves/production ratio for natural energy sources are:

Reserves/production ratio, world 1987
Oil	42 years
Natural gas	58 years
Coal	222 years

At the end of 1987, 417 nuclear power stations, with total capacity of about 300 GW(el), produced annually 16% of the total world consumption of electricity. In the next few years some further 100 GW(el) will be commissioned.

Some details concerning long-term resources are:

—oil (including heavy oils) and gas (including pressurized methane at a depth of 5–8 km) will satisfy the needs of mankind during almost the whole of the next century.

—coal (including brown coal, etc.) is present in amounts which will fulfil needs for two or three centuries.

—nuclear fission, in spite of all its difficulties and dangers, relies on uranium and thorium, of which there are copious amounts in the outer layer of the crust, making it possible to generate for millions of years. The indispensable condition for further development of this technique is, of course, a radical change in the design of power reactors in the direction of inherent safety and a positive solution of the long-term storage of radioactive wastes. It seems that there exist no physical, technical or even economic obstacles to achieving this state of safety within the next two or three decades.

—nuclear fusion (thermonuclear power reactors) has still not achieved technical fruition to allow prediction to be made for the next four to five decades. It seems that no physical or technical obstacles exist which will prevent thermonuclear power generation. Much more open to discussion is the question of its future economic competitiveness. The quantity of the appropriate energy carriers—deuterium and lithium (as breeding material for the production of tritium)—are of an order of magnitude corresponding to millions of years supply for future terrestrial civilisations, even if the energy consumption per capita becomes significantly larger.

—tapping the heat of the internal layers of the Earth seems to be only of local signifiance and cannot be the solution for the global energetic problem [93, 94].

—the use of solar energy indirectly through:
  wind
  oceanic heat
  kinetic energy of oceanic currents

—the technology of harnessing solar radiation in thermal and photovoltaic power stations is now fully ripe for exploitation, having only a rather small, though not negligible, economic handicap. The future use of solar power stations placed in stationary orbits around the Earth is far from being economically viable, but is not prohibitively expensive. The use of indirect solar energy, in the form of wind, oceanic heat, and oceanic currents, seems to be technically feasible but still economically handicapped. It is obvious that the application of solar energetics is the 'solution for eternity'.

From this point of view the problems of the future of mankind on this planet are not limited by the amount and quality of energy available, not even by environmental difficulties, assuming that mankind can halt the global population increase at a level of 8–9 giga people, and that the energy requirement per capita does not exceed three to five times the present level of about 2.3 kW per capita. This means that the present energy flow in the technosphere of some 10.5 TW will increase to around 60 to 80 TW (Fig. 35).

Another question is the direction of the use of energy. Parallel to the present major areas of consumption, such as housekeeping, industry, agriculture, transportation, etc., some new areas will grow in importance, such as material recycling, waste management, climate regulation and restoration of 'old wounds' in the natural environment.

Renewable Energy Resources

The undisturbed extraterrestrial solar radiation flux equals 1367 W/m², the so-called 'solar constant'. This value is known to an accuracy of 0.5%. It should be

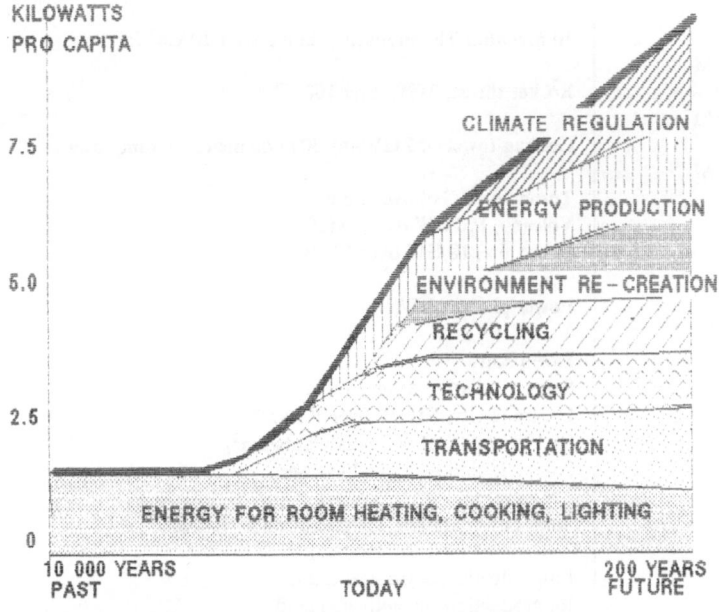

Fig. 35. Primary energy flux used in the future per capita

said that the true value varies from 1413 W/m² at the perihelion (the closest approach to the Sun) to 1321 W/m² at the aphelion. Since the cross sectional area of Earth to the surface of the sphere equals 4, the total solar energy flux on the Earth's outer atmosphere equals, on average, 1367/4 = 341.7 W/m². Taking into account the albedo (see Tables 32) of 0.31, the Earth retains only 240 W/m². However, because of absorption in the atmosphere, the average solar energy arriving at the Earth's surface which can be used by photosynthesis is 170 W/m² over the whole year; that is during $3.15 \cdot 10^7$ seconds per year. This corresponds to an energy of 170 Watt-year/ (year·square metre) or 5.35 GJ per year. The maximum solar energy flux, at noon reaches 1000 W/m² (Table 37).

Energy Perspectives for the Very Distant Future

Let us assume that:

—the present global population of 5.1 giga people will increase to 8–9 giga during the next one or two centuries and then maintain equilibrium for a very long time, and

—the energy flow in the technosphere which equals at present 2.3 kW per capita, will increase by four to five times and reach some 8 to 10 kW per capita.

Table 37. Energetics and the environment

Energy flux Watt/m²	
	10 Megaton Thermonuclear bomb over 10 km² for 1 s
100 M	
	Rocket thrust, 2 GW over 100 m² for 100 s
10 M	
	Cooling tower of 2 GW(th), 30 m diameter, continuously
1 M	
	Tornado, 50 GW over 1 km²
	Volcano, 100 GW over 1 km²
	Forest fire, 10 GW over 1 km²
100 k	
	Power plant, 1 GW(el)
10 k	
	Steel plant
1 k	Solar irradiation at maximum
	Centre of city, 50 GW over 100 km²
	(10 kW/capita, 50 000 inhabitants/km²)
	Solar irradiation, annual average
100	
	Solar power station, thermal or photovoltaic
10	
	Bioproduction of tropical forest
	Bioproduction on cultivated land
1	

The resulting man-made energy in the distant future will reach a level of 60 to 80 TW. This value will be used in further considerations of the impact on the natural environment.

The Earth has sufficient sources of free energy to fulfil the future demand of a highly developed society for a very long time, over mega or even giga years. The potential energy sources are the Sun, the fissionable nuclides U and Th, and the fusionable nuclides deuterium and tritium (bred from lithium). In the distant future it cannot be excluded that the 'import' of deuterium from Jupiter will be feasible.

In spite of the fact that there is no need for new and more exotic energy sources, from time to time such ideas are published. A repeated theme is the use of the energy of a vacuum. A number of quasi-scientific publications, and even some 'experiments', have tried to promote the discovery of such an 'eternal and free' energy source. Recent news of such a find has even come from a well-known Californian institute, published in one of the best scientific journals. To the question: "What constraints do the laws of physics place on the activities of an arbitrarily advanced civilization?" was given the answer: "One can imagine an advanced civilization pulling a wormhole out of the quantum foam and enlarging it to classical size." (A wormhole is a quantum-mechanical concept of the geometry of time-space at the level of size of 10^{-43} m and of time of 10^{-35} s). It was also proposed that the same phenomenon could allow, "Advanced beings . . . to travel backwards in time and" perhaps violate causality. My personal point of view is much more conservative and tends to refute such ideas as rather farfetched.

The most imporant conclusions from our point of view is that, at present, mankind has sufficient knowledge and the technological means to extract free energy from non-exotic, obvious sources at the appropriate level over the next mega- and gigayears. Of course, the environmental influence of waste thermal energy, radioactive wastes, and other dangerous wasteproducts, such as NOx, CO_2, etc., must be solved within the near future, even if the price is relatively high. However, here are well-based hopes that the price for the protection of the environment will not be prohibitively high.

Global Thermal Pollution

Let us first of all define two well-known quantities to make an evaluation of this subject easier. Firstly, the total solar radiation effectively absorbed by the Earth is close to 120 500 TW. This value has been determined to an accuracy of 0.5%. Let us then assume that the accuracy will reach 0.1%, giving an uncertainty of about 120 TW, caused not only because of difficulties in operational measurement of the solar constant, but also because of instability of the Sun, as a star, of the variation of the albedo, and for natural or man-made reasons.

Secondly, the present energy flow inside the man-made technosphere equals approximately 10.5 TW and is partially directly connected with the influx of solar energy through, for example hydropower, windpower, wood production, and agricultural wastes. The effective additional energy coming from the burning of fossil and nuclear fuels is estimated to be about 9 TW.

This man-made additional 9 TW must be taken in relation to the total terrestrial energy generation of 120 500 TW. This means that the present man-made energy production reaches a level of some 75 ppM. However, the man-made energy generation rate is only 7.5 percent of the uncertainty in the value of the total solar energy flux received, and as such cannot be considered as very important.

Assuming a foresighted approach concerning future energy production, in a highly developed human society in equilibrium in the very distant future the energy generation will reach some 60 to 80 TW, which is 6 to 8 times more than the present value. Even if the whole of this man-made generation would be decoupled from solar energy, such as through nuclear energy, the relation to the total terrestrial energy flux and even to its instabilities or uncertainties is very small and remains almost negligible.

From the values of Table 37 it is obvious that the local impact of the dissipation of man-made energy, decoupled from solar energy, is significant and often disastrous.

The cost of different forms of energy differs considerably, and can be seen in broad terms in Table 38.

Very Distant Future of Mankind: Mega and Giga Years

Distant Future of Mankind: Large Scale Factors

Previous evaluation of conditions for the further development of mankind on this planet ended rather optimistically. There are rather good possibilities that the

Table 38. Price of energy; human and technological

Prices in US$ (1988)
pro 1 Gigajoule of energy

100 000	Human intelligent work	Brain: 20 Watt, Salary $15/hour
10 000	Human work	Muscle: 40 Watt, $2.5/hour
1 000		
100	Human food	Contents: 2400 kcal/day = 10 MJ/day Price of food: $10/day
10	Human food	Contents: 2400 kcal/day = 10 MJ/day Price of food: $1/day
1	Electricity Crude oil Coal	$3/thousand kilowatt-hours $20/barrel $50/ton, mine head

following important problems can be positively solved:
—stabilisation of global population
—production of food
—recycling of matter
—production of energy
—limitation and even significant decrease of previous environmental scars
—reconstruction of the natural environment.

If all these problems are really solvable then there emerges the principal question of what chance has mankind of existing on this planet into the very distant future, measured not only by centuries or millenia but by mega or even by giga years. Of course, at each point in the evolution of mankind, foresight will change and will have peculiarities and even singularities. But this cannot prevent us from trying to formulate the best possible answer, even if its validity is very short. However, what can we say at present about the large scale factors ultimately governing our future?

Stability of Universe, Galaxies and the Sun

The stability of the Universe results from the stability, or better, from the constancy of natural laws and constants. The common opinion of physicists is that the set of known laws of Nature and the set of natural constants does not vary during the evolution of the Universe. This does not mean that the Universe itself is eternal. Not at all. Some hypotheses and physical models assume that the Universe, which at present is 15 gigayears old, will end in a very hot Anti-Big-Bang, the so-called Big Crunch, which will occur in about 100 gigayears. Another hypothesis predicts a 'cold end' to the Universe. In this model the Universe will exist for an even longer time, hundreds of orders of magnitude greater than the previous 100 gigayears. (see Fig. 13)

The question of the stability of the galaxies is not so exact, but the existence of the Milky Way is assured for at least a hundred gigayears.

Not so optimistic an answer must be given when the future of our central star, the Sun, is predicted. The Sun, as a typical 'main sequence' star, has a lifetime limited to the period during which part, but far from all, of its hydrogen will be burned; that is, transformed into helium. A relatively quiet period of some 5 gigayears lies before us. Later the Sun will, in a relatively short time, transform itself into a 'red giant', which will result in an increase of the terrestrial solar constant by more than one order of magnitude. The impact on thermal conditions on the Earth will be unequivocably fatal for the atmosphere, hydrosphere and, first of all, for the biosphere. There is no chance for existence of mankind under these conditions. Only a radical solution on a planetary or galactic scale could give some chance of survival. It may be that Nature does not prohibit salvation from this fate, but speculation about such a solution is at present unreasonable. [95]

Stability of Planet Earth

Is our planet stable enough to allow it to wait for a catastrophe which will come in gigayears? How stable is the planetary system, the rotational speed of the Earth

Table 39. Possible number of cosmic civilisations. All numbers are somewhat arbitrary, only given for the sake of illustrating our a priori assumption concerning the extremely small probability of success in the Search for Extraterrestrial Intelligence

	Stars capable	Stars incapable
All stars in Galaxy	130 giga	–
Stars far from Center	10 giga	120 giga
Stars nonmultiple	1 giga	9 giga
Stars longlived	100 mega	900 mega
Stars not too cold	10 mega	90 mega
Stars 3d generation	1 mega	9 mega
Stars with planets	100 kilo	900 kilo
Planets inside ecosphere	10 kilo	90 kilo
Planet with life	1 kilo	9 kilo
Planet with intelligence	100	900
Intelligence with cosmic techniques	10	90
Intelligence with cosmic interest	2[a]	8

[a] This means our civilisation and maybe one other within 'cosmic interest'. The mean separation distance in such a case equals 30 000 light years. Maximum mutual messages: 33 in a megayear.

and the global climate? Given that the Earth itself is relatively stable as a cosmic object, how stable are the constituents of our planet.

—the solid lithosphere, affected by continental drift and by erosion, by volcanic activity and by man,

—the atmosphere, a product of the activity of the biosphere, also affected by man,

—the hydrosphere, which is continuously destroyed by photolysis, depleted by escaping hydrogen, and influenced by falling icy cosmic bodies,

—the biosphere, the most active planetary factor?

The Earth as a whole is influenced by the impact of large cosmic bodies, such as comets, meteorites, and asteroids. The global climate is, on a long-term scale influenced not only by man-made substances, but also by the continuous drift of continents (the Atlantic ocean is only 150 million years old), by oceanic currents (the Gulf stream is relatively young), by its position relative to the Sun (the so-called Milankovitch theory), by the impact of large cosmic bodies, and last but not least, by the biosphere, which in the future can be manipulated by man in such a direction which can positively influence the global climate.

Extraterrestrial Connections

The ultimate factor influencing the future of mankind on this planet is, of course, Man himself. The further development of his intellectual abilities, his intelligence, his self-consciousness, and his spiritual state are deciding factors. Among other things, the possibility of meeting extraterrestrial intelligent life can have enormous, positive or negative, impact on the fate of mankind. The present state of science allows us to assess the possibility of meeting such an ETI (extraterrestrial in-

Abbreviations, Symbols

Symbol	Unit	multiply by	to achieve
Mass			
Pg	Petagram	1	10^{15} g
		1	10^{12} kg
		1	10^9 ton
ppM	part per Mega	10^6	1
ppG	part per Giga	10^9	1
ppT	part per Tera	10^{12}	1
ppP	part per Peta	10^{15}	1
Time			
y	year		
Myr	Megayear	10^{-6}	y
Gyr	Gigayear	10^{-9}	y
Surface			
Mm²	Squaremegametre	1	10^{12} m²
		1	10^6 km²
Temperature			
MKelvin	Megakelvin	1	10^6 Kelvin
GKelvin	Gigakelvin	1	10^9 Kelvin
Energy			
eV	Electronvolt	11605	Kelvin/k
MeV	Megaelectronvolt	10^{-6}	eV
GeV	Gigaelectronvolt	10^{-9}	eV
k	Boltzmann's constant	$8.615 \cdot 10^{-5}$	eV/K
eV/atom	Electronvolt per atom	96.484	kJ/mol
TWyear	Terawattyear	1	approx. $35 \cdot 10^{18}$ J
Power			
W(el)	Watt (electricity)		

telligence). The likelihood depends to a first approximation on the possible number of planets able to sustain life, and especially intelligent life. Table 39 gives some idea of the orders of magnitude of the realisation of the appropriate conditions for the emergence of an intelligent society elsewhere in our galaxy. Contact with extragalactic partners seems to be of probability lower by more than another order of magnitude (Table 39).

It should be noted that esoteric studies concerning the possibility of the emergence of a planetary biosphere and especially of the existence of a technical civilisation on a distant planet has some significance for terrestrial environmental science. Such studies are going on parallel to the effort to evolve a general, Earth-independent definition of life, intelligence and technical civilisation and to construct a model of a planetary environment which enables these to exist. [96, 97, 98, 99]

References

1. Davies PCW, Brown J (1988) Superstring. Cambridge Un. Press, Cambridge, New York
2. Schwarz JH (1987) Phys. Today, 11, 33
3. Green MB, (1986) Sci. Am. 255, (3) 44
4. Rozental I (1980) Sov. Phys. Usp. 23, 6, 296
5. Hawking SW (1987) The origin of the Universe. Lecture at Swiss Fed. Ins. Techn. Zürich
6. Linde A (1987) Phys. Today 9: 61
7. Abbott LF, So-Young Pi (1986) Inflationary Universe. World Sci. Pub., Philadelphia
8. Gunzig E, Geheniau J, Prigogine I (1987) Nature, 330, 621
9. Quigg C (1985) Sci. Am. 252, (4) 64
10. Taube M (1985) Evolution of matter and energy. Springer, New York, Heidelberg, Tokyo
11. Taube M (1988) Evolution der Materie und Energie und die Zukunft des Menschen. Hirzel, Stuttgart
12. Taube M (1982) Atomkernen. 40, 128
13. Taube M (1982) A Atomkernen. 40, 208
14. Novikow ID (1983) Evolution of the Universe. Cambridge Un. Pr. Cambridge
15. Burns JO (1986) Sci. Am. 255, 30
    Dressler A (1987) Sci. Am. 257, 38
16. Silk J (1987) Phys. Today, 4, 28
17. Gehrz RD, Black DC, Salomon PM (1984) Science, 224, 823
18. Chandrasekhar S (1984) Science 226, 497
19. Fowler WA (1984) Science, 226, 922
20. Bethe HA, Brown G (1985) Sci. Am. 252, 40
21. Helfand, D (1987) Phys. Today, 8, 25
22. Rank DM et al (1988) Nature 331, 505
23. Burrows A (1987) Phys. Today. 8, 28
24. Viola VE, Mathews GJ (1987) Sci. Am. 256, 34
25. Allen CW (1973) Astrophysical Quantities, 3rd ed. Athl Press, London
26. Landolt-Bernstein, Band 2, a,b, (1982) Springer, Berlin-Heidelberg-New York
27. Dyson F (1979) Rev. Mod. Phys. 51, 447
28. Islam JN (1979) Vistas in Astr. 23: 265
29. Balley JG (1986) Science 232: 185
30. Bailey ME, Clube SVM, Napier WM (1986) Vistas in astronomy 29: 53
    Balsiger H, Fechtig H, Geiss J (1988) Sci. Am. 259: 62
31. Stern SA, Shull JM (1988) Nature 332: 407
32. Kissel J, Krueger FR (1987) Nature 326: 755
33. Mitchell DL, Lin RP, Anderson KA, Carlson CW, Curtis DW, Korth A, Reme H, Sauvaud JA, d'Uston C, Mendis DA (1987) Science 237: 626
34. Buehler RW (1988) Meteorite. Birkhaeuser, Basel
35. Holland HD, Trendall AF (eds) (1984) Patterns of change in the Earth's evolution. Springer, Berlin Heidelberg New York
36. Siever R (1983) Sci. Am. 249: 30
    Jeanloz R(1983) Sci. Am. 249: 40
    McKenzie DP (1983) Sci. Am. 249: 50
    Franchetaeau J (1983) Sci. Am. 249: 68
    Burchfiel BC (1983) Sci. Am. 249: 86
37. Oxburgh ER, O'Nions RK (1987) Science 239: 1583
38. Press F, Siever R (1978) Earth, 2nd edn, Freeman, S. Francisco
39. Holland HD, Lazar B, McCaffrey M (1986) Nature 320: 27
40. Duxbury AC, Duxbury A (1984) The world's oceans. Addison-Wesley, Reading, MA
41. Chyba FC (1987) Nature 330: 632
42. Stanley SM (1986) Earth and life through time. Freeman, New York
43. Nance RD, Worsley TR, Moody JB (1988): Sci. Am. 44

44. Wood JA, Chang S (eds) (1985) The cosmic history of the biogenic elements and compounds. NASA series, SP-476, Washington, DC
45. Taube M (1964) Hydrogen as carrier of life, Nucl. Pub. House, Warsaw
46. Taube M (1971) Simple model of a living thing. Proc. First Europ. Biophys. Wiener Med. Akad. Verl. Vienna
47. Eigen M (1988) Stufen zum Leben. Piper, München
48. Wald G (1988) Science: 240, 2
49. Harold FM (1986) The vital force; a study of bioenergetics. Freeman, New York
50. Cairns-Smith AG (1985) Sci. Am. 256: 74
51. Schidlowski M (1988) Nature 333: 313
52. Luria SE, Gould SJ, Singer S (1981) A view of life. Benjamin/Cummins, Menlo Park, CA
53. May MR (1988) Science 241: 1441
54. Vidal G (1984) Sci. Am. 250: 32
55. Alvarez LW (1987) Phys. Today 7: 24
56. Officer CB, Hallam A, Drake CL, Devine JD (1987) Nature 326: 143
57. Hallam A (1987) Science, 238: 1237
    Wolbach WS, Gilmour I, Anders E, Orth CJ, Brooks RR (1988) Nature 334: 665
58. Crowley TJ, North GR (1988) Science 240: 996
59. Sepkowski JJ (1988) Extinctions of life. Los Alamos science: 36
60. Budyko M (1986) The evolution of the biosphere. Reidel, Amsterdam
61. Pastor J, Post WM (1988) Nature 334: 55
62. Schneider SH, Londer R (1984) The coevolution of climate and life. Sierra Club Books, San Francisco
63. Vitousek PM, Ehrlich PR, Ehrlich AH, Matson PA (1986) BioScience, 36: 368
64. Smil V (1983) Biomass energy. Plenum Press, New York
65. Detwiler RP, Hall CAS (1988) Science 239: 42
66. Platt T, Sathyendranath S (1988) Science 241: 1613
    Lorenz SE, Arnone RA, Wiesenburg DA, DePalma IP (1988) Nature 33: 245
67. World Resources 1986 (1986) Basic Books, New York
68. Keyfitz N (1984) Sci. Am. 250: 22
69. Gabor D (1981) Beyond the Age of waste. Pergamon Press, Oxford
70. Larson ED, Ross MH, Williams RH (1986) Sci. Am. 254: 24
71. Skinner BJ (1979) Proc. Nat. Acad. Sci. USA 76: 4212
72. Mayer C (1985) Science, 227: 1421
73. Goeller HE, Weinberg AM (1976) Science 191: 683
    Larson ED, Ross MH, Williams RH (1986) Sci. Am. 254: 24
74. Duckham AN et all (1976) Food production and consumption, North Holland, Amsterdam
75. Hinman CW (1986) Sci. Am. 255: 25
76. Spurr SH (1979) Sci. Am. 240: 62
77. Jensen NF (1978) Science 210: 317
78. Revelle R (1974) Sci. Am. 231: 160
79. Smil V (1985) Sci. Am. 253: 104
80. Vietmeyer ND (1986) Science 232: 1379
81. Kasting JF, Toon OB, Pollack JB (1988) Sci. Am. 258: 46
82. Peixoto JP, Oort AH (1984) Rev. Mod. Phys. 56: 365
83. Budyko MI (1982) The Earths climate: past and future. Academic, New York
    Nance D, Worsley TR, Moody JB (1988) Sci. Am. 259: 44
84. Covey C (1984) Sci. Am. 250: 42
85. Schmetz J, Raschke E (1986) Spekt. Wissen. 1: 96
86. Sagan C, Toon OB, Pollack JB (1979) Science 206: 1363
87. Schneider SH (1987) Sci. Am. 256, 72
88. World Energy Conference (1986) Proceed. 13th congress, Cannes
89. Fettweis GB (1979) World coal resources. Elsevier, Amsterdam
90. Deffes KS, MacGregor IO (1980) Sci. Am. 242: 50

91. McDonald A (ed) (1981) Energy in a finite World.
    Int. Inst. Appl. Syst. Anal. Laxenburg
92. Taube M (1974) Plutonium, Verlag, Chemie, Weinheim
93. Penney TR, Bharathan D (1987) Sci. Am. 256: 74
94. Moretti PM, Divone LV (1986) Sci. Am. 254: 88
95. Taube M (1981) J. Br. Interplan. Soc. 35: 219
96. Papiagannis MD (1985) Nature 318: 135
97. Taube M (1989) Menschliche Intelligenz (in press)
98. Martin AR (1986) J. Br. Interplan. Soc. 38: 276
99. Drake F (1988) Los Alamos Science 16: 50

Subject Index

Albedo, global energy flow 166
Alpha decay 78
Antimatter 78
Atmosphere 22
–, origin and evolution 112
Atomic bomb tests, contamination 56
Automobile emissions 29, 56
Autotrophy 130

Beryllium isotopes 17
Beta decay 74
Binding energy 75–77
Biomass 23
Biosphere 22
–, chemical composition 127
–, human use 135
Black holes 85, 78, 96
Bosons 69
Bricks 73

Carbon, nucleosynthesis 81, 84
Cement factories 50
Chernobyl disaster, radioactive isotopes 57, 58
Coal, elements in 51
Collision zones 19, 20
Cosmic matter 94, 97
– rays, nucleosynthesis 86
Cyprus ophiolite 13

Deuterium, primordial synthesis 79
Deuteron 74
Dinosaurs, extinction 133

Earth, energy 4, 5
–, structure 3
Earthquakes, environmental impact 132
Electrons 70
Elementary particles 70, 79
Elements, abundance 92
–, geochemical characterization 30
–, natural concentrations 29–33
–, soil, limits/guidelines 43–46

–, mobility 58
–, terrestrial abundance 105
Emissions, automobile 29, 56
Energy flow 164–173
– sources, non-renewable 169
– –, renewable 172
Eobionts 128
Evolution, chemical 97, 109, 113
Extinctions 131
Extraterrestrial civilizations 178

Fermions 69
Fertilizer, contamination 54
Fission, heavy atomic nuclei 78
Fluorine emitting industries 53
Food, synthetic 162
Fossil fuels 155, 170
Fungi, energy source 131
Fusion, light atomic nuclei 78

Galapagos ridge 10
Galaxies 80
–, stability 177
Gas-dust clouds 98
Geosphere mixing 1
Grand-unified force 69
Gravitational force 70, 71
Graviton 72

Hadrons 72
Half lives, radioactive elements 34, 35
Halley's comet 92, 99
Helium isotopes 21
–, nucleosynthesis 78
Heterotrophy 130
Himalayas 19
Hot spots 20
– –, magmatism 21
Hydrosphere 22
–, origin/evolution 112–115

Intergalactic matter 80

Juan de Fuca ridge   11

Life, origin   112, 113

Magma production   5
– –, compressibility   18
– –, ocean ridges   6, 7
– –, underplating   18
Marine productivity   137, 139
Matter, cycling   155
Mercury emitting industries   53
Metallurgical industry   47
Metals, nucleosynthesis   83
Meteorites   102, 103
–, amino acids   103
Milky Way, stability   177
Mining industry   47
Mortar   73

Neutrinos   73, 96
Neutron stars   85
Neutrons   74
Nitrogen, nucleosynthesis   81
–, origin of isotopes   83
Nucleosynthesis   77
Nuclides, binding energy   75

Ocean floor   12
– –, convection   12
– –, serpentinization   13
– ridges   5
– –, chemical processes   7
– –, magma production   6, 7
– –, strontium isotopes   10
– –, sulfate removal   9

Pathfinder elements, pedosphere   39
Pedosphere, geochemical exploration   37
Pest control agents, contamination   55
Photons,   72
Planetoids, accretion   105
Planets, stability   177
Pollution, thermal   175
Power plants, coal-fired   50
Primordial soup, origin of life   112
Productivity, continental/marine   139
Protobionts   128
Protoplanets   104
Protosun   90

Quarks   70–75

Radioactive elements   34, 56
Recycling, diamond/coesite   24

Repulsive forces   71
River flux   24
Russel-Hertzsprung diagram   81

Salton Sea   14
Serpentinization, ocean floor   13
Sewage sludge, contamination   54
Soils   23
–, anthropogenic contamination   29, 45
–, elements   38
–, horizons, elements   40
–, trace elements   35, 36
Stars, evolution   80
Subduction   14
–, crust   16
–, fluid transport   17
–, magmatism   17
–, ore deposits   18
–, sediments   15, 16
–, serpentine   14
– zones, magma production   17
Sun   88–91
–, origin/evolution   88
–, properties   91
–, red giant   91
Supernovae   85
Superstrings   68
Superunified force   69

Technotrophs   131
Terrestrial energy flow   111
Thallium contamination   49
Theory of everything   68
Three-alpha process   82
Tomography, seismic   3
Tritium, primordial synthesis   79
Tropical forests   136
Tunguska event   102, 114

Underplating, magma   18
Unified force   70
Universe, nucleosynthesis   77
–, open/closed   96
–, origin   69
–, stability   74
Uranium, binding energy   76
–, supernova   84
–, terrestrial energy source   110

Volcanic eruptions, climate   132

Water cooling   6, 7
Weathering   23
Wind transport   23

Volume 1 Series: The Natural Environment and the Biogeochemical Cycles

Volume 1A: The Atmosphere (M. Schidlowski)
            The Hydrosphere (J. Westall, W. Stumm)
            Chemical Oceanography (P.J. Wangersky)
            Chemical Aspects of Soil (E.A. Paul, P.M. Huang)
            The Oxygen Cycle (J.C.G. Walker)
            The Sulfur Cycle (A.J.B. Zehnder, S.H. Zinder)
            The Phosphorus Cycle (J. Emsley)
            Metal Cycles and Biological Methylation (P.J. Craig)
            Natural Organohalogen Compounds (D.J. Faulkner)

Volume 1B: Basic Concepts of Ecology (S.W.F. van der Ploeg)
            Natural Radionuclides in the Environment (R. Fukai,
            Y. Yokoyama)
            The Nitrogen Cycles (R. Söderlund, T. Rosswall)
            The Carbon Cycle (A.J.B. Zehnder)
            Molecular Organic Geochemistry (P.A. Schenck, J.W. de Leeuw)
            Radiation and Energy Transport in the Earth Atmosphere System
            (H.-J. Bolle)

Volume 1C: Humic Substances, Structural Aspects, and Photophysical, Photo-
            chemical and Free Radical Characteristics (G.G. Choudhry)
            Organic Material in Sea Water (P.J. Wangersky)
            Marine Gelbstoff (M. Ehrhardt)
            The Surface of the Ocean (L.W. Lion)
            Atmospheric Nitrogen. Chemistry, Nitrification, Denitrification,
            and their Interrelationships (R.D. Hauck)
            Carbon Dioxide: A Biogeochemical Portrait (E.T. Degens,
            S. Kempe, A. Spitzy)

Volume 1D: The Cycles of Copper, Silver and Gold (H.J.M. Bowen)
            Modelling the Global Carbon Cycle (G. Kratz)
            Chemical Limnology (T. Frevert)
            Environmental Microbiology (W.D. Grant, P.E. Long)

Volume 1E: The Thermodynamics of Ecosystems (L. Johnson)
            Environmental Systems (G.H. Dury)
            Global Transport Processes in the Atmosphere (J.R. Holton)
            The Atmosphere: Physcial Properties and Climate Change
            (R. Eiden)

Volume 2 Series: Reactions and Processes

Volume 2A: Transport and Transformation of Chemicals: A perspective
(G.L. Baughman, L.A. Burns)
Transport Processes in Air (J.W. Winchester)
Solubility, Partition Coefficients, Volatility, and Evaporation
Rates (D. Mackay)
Adsorption Processes in Soil (P.M. Huang)
Sedimentation Processes in the Sea (K. Kranck)
Chemical and Photo Oxidation (T. Mill)
Atmospheric Photochemistry (T.E. Graedel)
Photochemistry at Surfaces and Interphases (H. Parlar)
Microbial Metabolism (D.T. Gibson)
Plant Uptake, Transport and Metabolism (I.N. Morisson,
A.S. Cohen)
Metabolism and Distribution by Aquatic Animals (V. Zitko)
Laboratory Microecosystems (A.R. Isensee)
Reaction Types in the Environment (C.M. Menzie)

Volume 2B: Basic Principles of Environmental Photochemistry (A.A.M. Roof)
Experimental Approaches to Environmental Photochemistry
(R.G. Zepp)
Aquatic Photochemistry (A.A.M Roof)
Microbial Transformation Kinetics of Organic Compounds
(D.F. Paris, W.C. Steen, L.A. Burns)
Hydrophobic Interactions in the Aquatic Environment
(W.A. Bruggeman)
Interactions of Humic Substances with Environmental Chemicals
(G.G. Choudhry)
Complexing Effects on Behavior of Some Metals (K.A. Daum,
L.W. Newland)
The Disposition and Metabolism of Environmental Chemicals by
Mammalia (D.V. Parke)
Pharmacokinetic Models (R.H. Reitz, P.J. Gehring)

Volume 2C: OECD Fate and Mobility Test Methods (A.W. Klein)
Biodegradation and Transformation of Recalcitrant Compounds
(A.H. Neilson, A.-S. Allard, M. Remberger)
Biodegradation of Water-Soluble Compounds (H.A. Painter,
E.F. King)
The Fugacity Concept in Environmental Modelling (S. Paterson,
D. Mackay)

Volume 2D: Hydrology (R. Herrmann)
Outdoor Ponds: Their Construction, Management, and Use in

Experimental Ecotoxicology (N.O. Crossland, C.J.M. Wolff)
Hydrolysis of Organic Chemicals (Th. Mill, W. Mabey)
Exchange of Pollutants and Other Substances Between the
Atmosphere and the Oceans (M. Waldichuk)
Root-Soil Interactions (P.B. Tinker, P.B. Barraclough)
Reaction Types in the Environment (C.M. Menzie)

Volume 3 Series: Anthropogenic Compounds

Volume 3A: Mercury (G. Kaiser, G. Tölg)
Cadmium (U. Förstner)
Polycyclic Aromatic and Heteroaromatic Hydrocarbons
(M. Zander)
Fluorocarbons (J. Russow)
Chlorinated Paraffins (V. Zitko)
Chloroaromatic compounds Containing Oxygen (C. Rappe)
Organic Dyes and Pigments (E.A. Clarke, R. Anliker)
Inorganic Pigments (W. Funke)
Radioactive Substances (G.C. Butler, C. Hyslop)

Volume 3B: Lead (L.W. Newland, K.A. Daum)
Arsenic, Beryllium, Selenium and Vanadium (L.W. Newland)
C_1 and C_2 Halocarbons (C.R. Pearson)
Halogenated Aromatics (C.R. Pearson)
Volatile Aromatics (E. Merian, M. Zander)
Surfactants (K.J. Bock, H. Stache)

Volume 3C: Aromatic Amines (L. Fishbein)
Phosphate Esters (D.C.G. Muir)
Phthalic Acid Esters (C.S. Giam, E. Atlas, M.A. Powers, Jr.,
J.E. Leonard)
Thallium (J. Schoer)

Volume 3D: Cellulose Production Processes (C.C. Walden, D.J. McLeay,
A.B. McKague)
Asbestos (P.E. Ney)
Carbon Black (D. Rivin)
Creosote (G. Sundström, A. Larsson, M. Tarkpea)
Elemental Phosphorus (R.F. Addison)
Molybdenum (G.A. Parker)

Volume 3G: Isocyanates (F. Brochhagen)
Nitro Derivatives of Polycyclic Aromatic Hydrocarbons
(NO_2- PAH) (H. Fiedler, W. Mücke)

Chlorinated Ethanes: General Sources, Biological Effects, and Environmental Fate (J. Konietzko, K. Mross)
Organic Explosives and Related Compounds (D.H. Rosenblatt, E.P. Burrows, W.R. Mitchell, D.L. Parmer)

Volume 4 Series: Air Pollution

Volume 4A: Air Pollution in Perspective (A. Wint)
Halogenated Hydrocarbons in the Atmosphere (P. Fabian)
Formation, Transport and Control of Photochemical Smog (H. Güsten)
Atmospheric Distribution of Pollutants and Modelling of Air Pollution Dispersion (H. van Dop)
The Mathematical Characterization of Precipitation Scavenging and Precipitation Chemistry (J.M. Hales)

Volume 4B: Peroxyacyl Nitrates (Pans): Their Physical and Chemical Properties (J.S. Gaffney, N.A. Marley, E.W. Prestbo)
Semivolatile Organic Compounds in the Atmosphere (R. Harkov)
Arctic Haze (G.E. Shaw, M.A.K. Khalil)
Air Pollution and Materials Damage (F.W. Lipfert)
Air Pollution Control Equipment (H. Brauer)

Volume 4C: Lichens as Indicators of Air Pollution (T.H. Nash, C. Gries)
Morbidity Associated with Air Pollution (M. Lippmann)
Mortality and Air Pollution (F.W. Lipfert)

Volume 5 Series: Water Pollution

Volume 5A: Epidemiologic Studies of Organic Micropollutants in Drinking Water (G.F. Craun)
Water Quality Genesis and Disturbances of Natural Freshwaters (M. Falkenmark, B. Allard)
Eutrophication of Lakes, Rivers and Coastal Seas (H.L. Golterman, N.T. de Oude)
Mathematical Models for Describing Transport in the Unsaturated Zone of Soils (W.T. Piver, T. Lindström)

The Handbook of
Environmental
Chemistry

Edited by O. Hutzinger

Volume 3

N. T. de Oude, Strombeek-Bever, Belgium (Ed.)

Detergents

Part F

1992. Approx. 460 pp. 48 figs. 116 tabs. Hardcover DM 248,–
ISBN 3-540-53797-X

Contents: *P. Christophliemk, P. Gerike, M. Potokar:* Zeolites. –
J. S. Falcone, J. G. Blumberg: Anthropogenic Silicates. –
J. L. Hamelink: Silicones. – *D. Gleisberg:* Phosphate. –
W. E. Gledhill, T. Feijtel: Environmental Properties and Safety
Assessment of Organic Phosphonates Used for Detergent
and Water Treatment Application. – *H. L. Hoyt,*
H. L. Gewanter: Citrate. – *H.-J. Opgenorth:* Polymeric
Materials, Polycarboxylates. – *J. G. Batelaan,*
C. G. van Ginkel, F. Balk: CMC. –
P. A. Gilbert: EDTA and TAED. –
R. S. Boethling, D. G. Lynch: Quater-
nary Ammonium Surfactants. –
H. A. Painter: Anionic Surfactants. –
R. J. Watkinson, G. C. Mitchell,
M. S. Holt: The Environmental
Chemistry, Fate and Effects of
Nonionic Surfactants. – *K. Raymond,*
L. Butterwick: Perborate. –
H. B. Kramer: Fluorescent Whitening
Agents.

Springer-Verlag
Berlin
Heidelberg
New York
London
Paris
Tokyo
Hong Kong
Barcelona
Budapest

The Handbook of
Environmental Chemistry

Edited by O. Hutzinger

Volume 4

O. Hutzinger, University of Bayreuth, FRG
(Ed.)

Air Pollution

Part C

1991. XII, 185 pp. 46 figs. 10 tabs.
Hardcover DM 152,– ISBN 3-540-53999-9

Contents:

H. Nash, C. Gries: Lichens
as Indicators of Air
Pollution. – *M. Lippman:*
Morbidity Associated
with Air Pollution. –
F. W. Lipfert: Mortality
and Air Pollution.

Springer-Verlag
Berlin
Heidelberg
New York
London
Paris
Tokyo
Hong Kong
Barcelona
Budapest